ROCKS
AND
RICHES

ROCKS AND RICHES

EXPLORING CALIFORNIA'S STUNNING GEOLOGY

GARY L. PROST

ILLUSTRATIONS BY ADAM PROST

Heyday, Berkeley, California

Copyright © 2025 by Gary L. Prost
Illustrations (unless otherwise noted) copyright © 2025 by Adam Prost

All rights reserved. Except for brief passages quoted in a review, no portion of this work may be reproduced or transmitted in any form or by any means, electronic or mechanical, including photocopying and recording, or by any information storage or retrieval system, or be used in training generative artificial intelligence (AI) technologies or developing machine-learning language models, without permission in writing from Heyday.

Library of Congress Cataloging-in-Publication Data

Names: Prost, Gary L., 1951- author. | Prost, Adam, illustrator.
Title: Rocks and riches : exploring California's stunning geology / Gary L. Prost ; illustrations by Adam Prost.
Description: Berkeley, California : Heyday, [2025] | Includes bibliographical references.
Identifiers: LCCN 2024055980 (print) | LCCN 2024055981 (ebook) | ISBN 9781597146777 (paperback) | ISBN 9781597146784 (epub)
Subjects: LCSH: Geology--California--Popular works. | Geology, Economic--California--History--Popular works. | Geology--Social aspects--California--History--Popular works. | California--Historical geography--Popular works. | California--History--Popular works.
Classification: LCC QE89 .P758 2025 (print) | LCC QE89 (ebook) | DDC 557.94--dc23/eng/20250317
LC record available at https://lccn.loc.gov/2024055980
LC ebook record available at https://lccn.loc.gov/2024055981

Cover Art: Wayne Alcorn / National Park Service (public domain)
Cover Design: Archie Ferguson
Interior Design/Typesetting: Victor Mingovits

Published by Heyday
P.O. Box 9145, Berkeley, California 94709
(510) 549-3564
heydaybooks.com

Printed in East Peoria, Illinois, by Versa Press, Inc.

10 9 8 7 6 5 4 3 2 1

Contents

Preface ix

A Very Brief Introduction to California Geology 1
It's about Time 1
Jurassic Coast 3
Plate Tectonics Means Constant Change 7

Chapter 1: Coast Ranges 17
Marin Headlands 17
    STOP 1: PILLOW BASALTS AT POINT
        BONITA LIGHTHOUSE
    STOP 2: DEFORMED FRANCISCAN RIBBON CHERTS
Mount Tamalpais and San Francisco Bay 24
    STOP 3: ROCK SPRINGS TRAILHEAD SERPENTINITE
    STOP 4: MOUNT TAMALPAIS SUMMIT AND THE
        EVOLUTION OF SAN FRANCISCO BAY
The San Andreas Fault 32
    STOP 5: BEAR VALLEY VISITOR CENTER
        AND THE SAN ANDREAS FAULT
Point Reyes National Seashore 36
    STOP 6: POINT REYES LIGHTHOUSE
    STOP 7: DRAKES BEACH
Mayacamas Mountains 43
    STOP 8: PETRIFIED FOREST, CALISTOGA
    STOP 9: THE GEYSERS

The Sonoma Volcanics.. 51
    STOP 10: PALISADES OF THE SONOMA VOLCANICS
    STOP 11: GLASS MOUNTAIN
Napa Valley and the *Terroir* of Wine 55
    STOP 12: NAPA VALLEY AND QUINTESSA VINEYARDS

Chapter 2: The Great Valley 63

The Great Valley Sequence............................ 64
    STOP 13: THE GREAT VALLEY SEQUENCE
        AT MONTICELLO DAM
The Devil's Thicket...................................... 69
    STOP 14: MOUNT DIABLO

Chapter 3: Gold Rush Country 81

The Sierra Foothills..................................... 84
    STOP 15: EMPIRE MINE STATE HISTORIC PARK
    STOP 16: MALAKOFF DIGGINS STATE HISTORIC PARK
    STOP 17: PRE-SIERRA NEVADA STREAM
        GRAVELS, I-80 ROADCUT AT GOLD RUN
    STOP 18: SUTTER'S MILL AND MARSHALL GOLD
        DISCOVERY STATE HISTORIC PARK, COLOMA
Placerville... 104
    STOP 19: GOLD BUG PARK AND MINE
    STOP 20: EL DORADO COUNTY HISTORICAL MUSEUM

Chapter 4: Sierra Nevada 111

Yosemite National Park and Yosemite Valley 122
    STOP 21: TUNNEL VIEW
The High Country...................................... 131
    STOP 22: OLMSTED POINT AND TENAYA LAKE
    STOP 23: LEMBERT DOME AND
        TUOLUMNE MEADOWS

STOP 24: GLACIAL TILL
STOP 25: TIOGA PASS AND MOUNT DANA VIEW

Chapter 5: Basin-and-Range ... 143

The Eastern Slope ... 147
STOP 26: SIERRA NEVADA METAVOLCANIC ROOF PENDANT AT TIOGA LAKE
STOP 27: PRE-SIERRA NEVADA METASEDIMENTS, ELLERY LAKE
STOP 28: LATERAL MORAINE

Mono Lake and Lee Vining ... 154
STOP 29: MONO BASIN SCENIC AREA VISITOR CENTER
STOP 30: TUFA PINNACLES AT MONO LAKE

Bodie and Travertine Hot Springs ... 163
STOP 31: BODIE STATE HISTORIC PARK AND GHOST TOWN
STOP 32: TRAVERTINE HOT SPRINGS

The Long Valley Caldera and Sierra Escarpment ... 173
STOP 33: WILSON BUTTE, INYO CRATERS
STOP 34: SIERRA NEVADA FAULT VIEW

Mammoth Mountain and Devils Postpile National Monument ... 179
STOP 35: EARTHQUAKE FAULT
STOP 36: MAMMOTH MOUNTAIN
STOP 37: MINARET VISTA
STOP 38: DEVILS POSTPILE NATIONAL MONUMENT

Owens Valley ... 188
STOP 39: CONVICT LAKE
STOP 40: MCGEE CREEK SCARP
STOP 41: BISHOP TUFF AND VOLCANIC TABLELANDS, SHERWIN SUMMIT
STOP 42: BRISTLECONE PINES IN THE PATRIARCH GROVE, WHITE MOUNTAINS

STOP 43: OWENS VALLEY FAULT
STOP 44: MOVIE FLATS, ALABAMA HILLS
STOP 45: SIERRA NEVADA FAULT

Death Valley National Park 216

STOP 46: FATHER CROWLEY VISTA POINT
STOP 47: MOSAIC CANYON
STOP 48: MESQUITE FLATS SAND DUNES
STOP 49: FURNACE CREEK VISITOR CENTER
STOP 50: ZABRISKIE POINT
STOP 51: ARTIST'S PALETTE
STOP 52: DEVIL'S GOLF COURSE
STOP 53: BADWATER BASIN
STOP 54: MORMON POINT AND THE
 COPPER CANYON TURTLEBACK
STOP 55: SPLIT CINDER CONE
STOP 56: RESTING SPRING PASS

Epilogue 245
Acknowledgments 249
Glossary 251
Notes 273
Selected Bibliography 275
About the Author 279
A Note on Type 281

Preface

The gaggle of kids ranged from eight to ten years old. They had just learned about the California Gold Rush a hundred years before, and gold fever was in the air. Running home, they asked their mom or dad for a hammer. Gathering again in the graveled alley at the end of the block, they started breaking open the pebbles that were lying all around. Soon they were showing each other the "glitter" in the rocks. Was it really gold?

    I was one of those kids. Noticing my interest in rocks, an older boy in the neighborhood helped me start collecting. He gave me some real doozies: **quartz**,* **pumice**, even enchanting pebbles of rounded **obsidian** that look black but let light through. Later, when I was in the Boy Scouts, the Geology merit badge continued to pique my interest in rocks, and earning the badge was easy because I already had a rock collection. In high school I spent two summers working as a backcountry guide at a Scout camp in Sequoia National Park. There I learned how **glaciers** carved and polished the stunning **granite** landscape of the Sierra Nevada.

    All through high school I wanted to be a classical archaeologist. You know, Indiana Jones. During my senior year, Mr. Helwick, our civics teacher, asked us to do a term paper on our chosen profession. After having volunteered at the UCLA Archaeological Survey, I realized there was little chance I could make a living as an archaeologist, since the

---

* Terms in boldface can be found in the glossary at the end of the book.

competition was fierce and there was little demand outside of a few museums and universities. My dad suggested that I look into geology since I liked being outdoors, was interested in rocks, and hey, there should always be a need for resources. As an electrical engineer in the aerospace industry in 1960s Southern California, my dad had been laid off twice and didn't want his son to go into a career where there was the potential for layoffs. It made sense to me at the time, and I took his advice. I spent forty years exploring first for minerals, and later as a petroleum geologist. I only got laid off once. (During one of several industry downturns in my career, there was a popular quip among petroleum geologists: "What do you call a geologist who brings his lunch to work? An optimist.")

I am new to California, and old to California. I was born and raised in the Los Angeles area, but left to go to university and pursue a career as a wandering geologist. I've lived in Flagstaff, Denver, Houston, and Calgary, and worked on projects in thirty-plus countries on every continent except Antarctica. I didn't really return to California until I retired, when we moved back to the San Francisco Bay Area. Although I knew the general outlines of California geology, I didn't really know the details. I knew the San Andreas Fault was the western edge of the North American **tectonic plate**, but didn't know that the same rocks found near Salinas are found at Point Reyes. I had done a lot of hiking in the Sierra Nevada and knew that it was mostly granite, but I didn't know how old the granite was, or how recently the mountains had formed. How could I discover more about the place I now called home?

I decided to take a road trip across the state looking specifically at the landscape, with a view to understanding how it evolved. I've always experienced a sense of wonder when watching a sunset progress over the **Central Valley** from Moro Rock in Sequoia Park, or observing a rainbow in the mist of Vernal Falls in Yosemite. But I'm also curious about rocks—

just ask my wife about all the specimens lying around the house. For me the sense of awe is even more profound when I know more about what I'm looking at—for example, knowing that the Jurassic antecedents of the present-day Sierra Nevada were in essence the same as today's Andes, where the **Nazca Plate** is plunging under the **South American Plate**. Understanding that **ribbon cherts**, the thin-banded red-and-green strata I see on the Marin Headlands, are the residue of millions of tiny shells of **plankton**-like sea creatures that lived in a tropical sea thousands of miles out in the Pacific. It was fascinating to learn that the San Joaquin River used to drain Nevada, but was cut off by the uplift of the Sierra, both shortening the river and turning Nevada into a **Great Basin** with no outlet to the sea. I was amazed to learn that the source of much of the gold in California rivers lies somewhere in Nevada. Putting local geology into a regional context and comprehending the interplay of time and process, the manner of forming rivers and mountains—features that we consider to be eternal—adds a whole new depth and dimension to understanding a landscape. I wanted to learn more about the rocks, but also the riches: How did the resources—from gold that glitters to black gold, from the bountiful soils to the world-class wine and fine peaches that spring from those soils, from boron to turbine-turning steam—create this vibrant state?

There have been a lot of "on the road to see America" books, from de Tocqueville to Steinbeck, from Kerouac to Bryson. My inspirations are less philosophical travelogue and more naturalist scientific, books such as *Assembling California* by John McPhee, *Hard Road West* by Keith Meldahl, and *Deep Oakland* by Andrew Alden.

This is not meant to be a comprehensive account of California's geology. There are other books for that—for example, the two-volume Roadside Geology series on California and even a slightly more detailed pair of books I wrote called *North America's Natural Wonders*.[1] For this book, I wanted to traverse the state from the coast to the mountains to the desert, and visit iconic locales that influence the history and economy of the state—places such as the wine country, Yosemite, and Death Valley. I wanted to understand the geology and history of the Gold Rush that created the modern state, the devastating impact of the precious metal on the Indigenous peoples who had inhabited the land before the prospectors came, and the influence of geology on modern agriculture, energy extraction, and water resources. Geology contributed to the diversity of Native cultures that occupied the land for five hundred generations; geology determined that California would become a state after gold was discovered; geology made the state rich in oil and geothermal resources; geology gave the state its agricultural bounty. I wanted to see geology that had a role in each of these outcomes. Because the geology of California is divided into more or less north–south segments parallel to the coast, crossing it from west to east exposed me to all the geologic provinces. I could have gone elsewhere, but this transect seemed to hit many of the main highlights all in one go. I look forward to exploring Southern California and northernmost California another time.

It occurred to me that others might find such an exploration interesting as well. I've always kept a field notebook, and this time was no exception. In this journal I jot down my observations, describe what I'm taking a photo of, and ask questions like "What rocks are these?" and "How old are they?" Then, once I'm back at my computer, I can look up the geology and answer my questions.

It is possible to understand geology on several different levels, from the simple "What are those rocks, and how did they form the landscape?" to the more comprehensive "What was the depositional environment of those sediments?" and "What was the deformation history of those rocks?" I've always felt that comprehending how a landscape formed gives me a deeper sense of understanding its history, much like the way talking to a person over a cup of coffee or a beer gives a deeper appreciation of that person and their background than just glancing at them from across a room. This book is about you and me having a long conversation with California. You don't have to know anything about geology—I'll guide you.

Geology propelled and continues to drive the history of California. The rich soil and mild climate, and separation by mountain ranges, made warfare difficult and led to an abundance of food sources and raw materials for tools that allowed the original inhabitants to live a relatively peaceful existence before Europeans arrived. In other parts of the continent, the first Europeans came to build farms, establish ranches, or trap beavers for their pelts. It was the discovery of gold that brought miners to California and transformed it overnight from a lazy Spanish backwater to a bustling American frontier. Later, the discovery of oil brought another kind of prospector to develop the prolific fields of Los Angeles and the Central Valley. Agriculture was almost an afterthought, but there, too, geology played a crucial role in creating the fertile soil and the mountains to catch the clouds pregnant with moisture. Tectonic activity and volcanism along the Pacific–**North American Plate** margin made **geothermal energy** more abundant here than anywhere else in North America. What does it say about us that greed motivated the first American settlers, yet the aftereffects of their plundering of the land motivated the first environmentalists? Some of the most impactful environmental actions of modern times have taken place in California, both with John Muir's successful efforts to make Yosemite a national park and with the Santa Barbara oil spill galvanizing the state to act to save seabirds, marine mammals, and beaches. It was in Yosemite that North American geologists argued about the reality of **ice ages** and glaciation, and where that concept was largely accepted. It was in California that the concept of **plate tectonics**, a game changer in understanding world geology, found support through the discovery of subduction zones and **transform margins**. In *Deep Oakland*, Alden says with a flourish, "Our science is rich and intellectually sublime. The

xiv   Preface

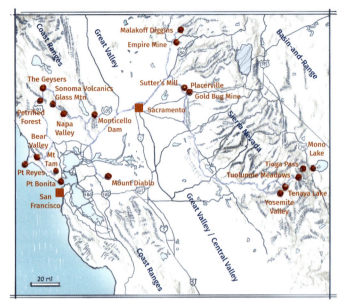

Western half of the California Geology Tour, with major stops

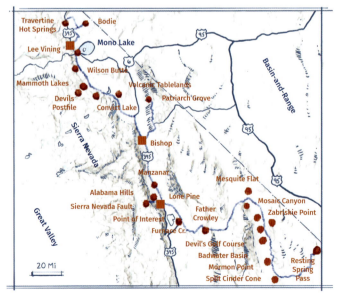

Eastern half of the California Geology Tour, with major stops

geologist looks for principles in the swarm of particularities that make up landscapes and regions. The point of finding the scientific laws of rocks and landscapes is to apply them, using those laws to gain insight into every place on Earth. The aim of geology is to help us awaken tomorrow and know our home ground afresh."[2] My purpose is to know my home ground afresh and to share this awareness with you.

From time to time, I will digress from geology to talk about history, culture, plants, and animals that are related to the area at hand. I don't apologize for this, as I find a holistic approach much more interesting.

My travels followed a rough plan, starting at the coast just north of San Francisco, crossing the state in zigzag fashion and ending in Death Valley. As you are reading this, you may sense that it is a composite of several tours, and you would be correct. I've tried to weave them together into a coherent narrative. For parts of the trip, I was accompanied by my wife; for other parts, by my son and daughter-in-law; in still other sections, I was on my own. I learned there are advantages to having a good navigator and, when the rocks are abundant and interesting, someone to keep you from running off the road.

Heading east, I examined first the Coast Ranges, then the Central Valley, then the Gold Rush country of the Sierra foothills. I wanted to know more about Yosemite, the jewel in the crown of California's national parks, and the mountains John Muir called the Range of Light. Crossing the divide to the eastern, arid part of California, I entered the province known to geologists as the **Basin-and-Range**, a land of **fault**-bounded valleys and ranges that extends all the way to Salt Lake City and indicates that the Earth is extending, pulling apart. I ended my journey at Resting Spring Pass, between Death Valley and Las Vegas. Here, in the eastern Mojave Desert, the continent is stretching, being pulled apart to the extent that the floor of Death Valley has actually dropped hundreds of feet below sea level. And just 85 miles away is Mount Whitney, the highest point in the US outside of Alaska. This is a land of contrasts.

This is a personal trip journal and guidebook that I hope will inform and entertain you. I'm not writing this for geologists, although I hope they would find it interesting and useful. No, this was written for my family and friends, who either shrug or roll their eyes when I get into a technical geological explanation. This is something I would like them to understand and enjoy. I do use some geology jargon, both because the terms have specific meanings and because I don't want to talk down to you. I've attempted to define geological terms in everyday language and with examples, but I might have missed some, and I use some terms repeatedly throughout the book. For those, I include a glossary at the end as a quick and handy reference. Terms that are included in the glossary have been set in **boldface** at their first mention in the text. (Some of these terms have already appeared.)

You may notice that nowhere was I beset by rain, snow, or wildfire smoke. The first two are common during our short spring; the third is becoming increasingly common in our extended summer fire season. California really has only two seasons: a hot, dry summer, which lasts from May to October, and a cool, wet spring, which lasts from November to April. I bring this up because when it comes to road trips, as with most of life, I have found that luck and timing are everything. I did my trip during the cool season, but also got lucky: skies stayed blue and temperatures were comfortable.

Like the people who make up the state, California's geology is diverse, complex, and unique, and much of it came from somewhere else. I got to thinking, maybe other people would like to take a similar road trip, get out and see beautiful scenery, and learn a bit about the geology that underlies the landscape. How would I explain the rocks to someone who wasn't a geologist? I'd have to start by talking about time.

A Very Brief Introduction to California Geology

In talking about rocks, it is impossible to ignore their age. Most of us measure time in minutes or hours—how long it takes to commute to work, or when we have to pick up our kids from soccer practice. Maybe, once in a while, we think in terms of years—next year our oldest is off to university, or we plan to finally take that vacation to Hawaii. That's not going to cut it when we talk about the age of rocks.

IT'S ABOUT TIME

When it comes to dating rocks, geologists needed something beyond their everyday experience. They came up with the notion of "deep time." This refers to expanses of time that are beyond our daily lives, way beyond our lifetimes and those of our parents and grandparents. For most of us it is incomprehensible that northern Europe and North America were buried under miles of ice just 20,000 years ago. So to say that a rock layer is 230 *million* years old defies understanding. But we know for a fact that the Earth is close to 4.6 billion years old, and the oldest rocks with visible fossils in them are around 530 million years old. These ages can be measured in a lab using the ratios of **radioactive isotopes** of uranium and thorium, which we know decay at a given rate.

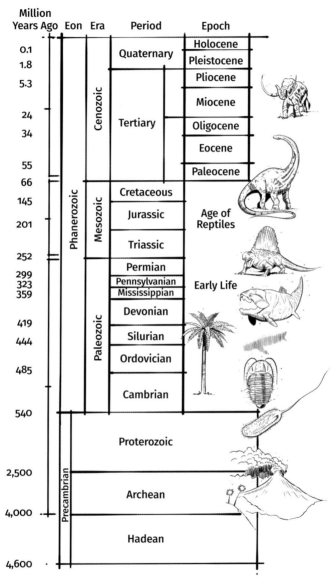

The geologic time scale. Most of the rocks in California are Jurassic and younger—that is, less than 200 million years old, although some found in Death Valley are up to 1.7 billion years old.

Most of geology deals with time in millions of years. To put this kind of time into perspective, a million years is to the age of the Earth as 19 seconds is to a day. *Homo sapiens* (our species) has been around for 250,000 years; if the age of the Earth were compressed into a year, then the time of humans would be the last 22 minutes of December 31.

To put this at a human scale, a generation is the time from your birth to the birth of your child, and averages around 25 years. So there are about 40 generations in a thousand years and 40,000 generations in a million years. Here's another way to think about it: let's say that the history of the Earth is written in a really big book where each page covers a million years. The book would be 4,600 pages long. The first fossils don't show up until around page 4000, and the first animals don't crawl out of the sea until 525 pages from the end. The dinosaurs are described on pages 4250–4535. Our human ancestors (the genus *Homo*) evolved 2½ pages from the end, and our species occupies just the last quarter of the last page. This kind of boundless time is the currency geologists deal in.

Once you wrap your head around time, the rocks are easy.

JURASSIC COAST

Point Bonita is dying. Grain by tiny sand grain, it is falling into the sea. It is the last of the continent poking into the Pacific, San Francisco's very own Land's End. And the ocean is winning, relentlessly wearing it away.

The first time I visited Point Bonita on the Marin Headlands, I was swept away by the beauty of the location. Rocky crags drop precipitously to churning, foaming waves bashing themselves to oblivion against the rising rocks. The water is cold and dark, a turbid green where it heaves itself against the sea cliffs with a muffled roar. There is a smell of seaweed and salt, and the sunlight shimmers in the hazy air. Ocean fog, the "marine layer" that forms when warm air moves over cold water, glides silently into San Francisco Bay.

A lighthouse, built to guide ships safely through the Golden Gate, is perched precariously at the end of a narrow ridge of rock jutting into the sea, obviously vulnerable but apparently oblivious to the crumbling cliffs. Some rocks are rounded and green; others are banded and contorted. The rocks here are varied, and the layers, where they exist, don't extend very far. It's not like other places in the western United States where a rock layer of a particular age can be followed for hundreds of miles and from state to state. That type of geology has been likened to a layer cake. No, here the rocks are discontinuous, almost like a fruit cake or plum pudding, with different types of rock all lumped together. The geology of California's coast is unique, and it took some new theories of the Earth to figure out why.

The time is the Jurassic, roughly 150–200 million years ago. Somewhere in the middle of the vast Pacific, in the inky depths, a miles-long crack opened in the ocean floor. Black, ropy lava inched upward through the crack, driven by the Earth's pressure, squeezed out of a molten pool not too far below. As the lava hit the cold, briny water, it formed pillows, red-hot blobs that hardened from the outside inward. The pillow-shaped blobs of lava, or **pillow lava**, piled up on older lava that had formed the same way. This is how ocean floor is made—slowly, incrementally, adding to itself over eons.

Driven by the heat of **radioactive decay**, currents of molten rock deep in the Earth move upward, forming a ridge of hot young rock. The heat dissipates in the frigid depths of the ocean. The crust on either side of the upwelling magma is carried away from the **mid-ocean ridge** by the circulating currents of molten rock below. The new ocean floor thus formed at mid-ocean **spreading centers** grows at about the same rate as fingernails, an inch or two per year.

Giant marine reptiles, dagger toothed and the size of whales, and coiled **ammonites**, a distant relative of the nautilus, dominated the oceans. These monsters shared their domain with another group, much less impressive but vastly more numerous: free-floating, single-celled plankton that lived near the surface. By the billions these microscopic animals grew, as they still do today, in and around tiny shells made of **silica** or **lime**. When these creatures die, their shells rain down on the ocean floor. **Radiolarians** are one variety of plankton with a glassy, silica shell that can resemble anything from a spiky ball to a bell to a pockmarked star. During the Jurassic, accumulations of dead radiolarians formed layers of **siliceous ooze** an inch or two thick on the deep ocean floor. These layers were separated by even thinner layers of marine mud. Marine mud accumulates very slowly as a result of dust falling out of the atmosphere and traces of **volcanic ash** drifting down from far-away eruptions. Eventually, after millions of years and burial beneath other sediments, the radiolarian layers hardened into ribbons of **chert**, a hard, glassy rock related to flint. The interlayered marine mud became mudstone.

At the same time as mud was accumulating in the deep ocean, sediments brought to the coast by rivers big and small piled up in unstable wedges just offshore from beaches. Every now and then the pile got too large and collapsed, or an earthquake gave the sediments a good shake. When that happened, the sediment pile flowed down the continental slope into the deep ocean.

A slurry of sediment and water, called a **turbidity current**, can flow tens or hundreds of miles out into the ocean before settling out of suspension. The big grains, being the heaviest, settle out first, followed in turn by medium, small, and then very fine particles. A coarse-to-fine graded deposit like this is known as a **turbidite**, and these are found in all the deep oceans around continents. The dirty sandstones that make up a turbidite are called **graywacke**. The name comes from the old German words for "gray stone."

The discovery of turbidity currents is a great example of geologic sleuthing. On November 18, 1929, a **magnitude** 7.2 earthquake struck off the coast of Newfoundland, Canada.

The shaking lasted an epic five minutes in a part of the world where earthquakes are virtually unknown. The quake destabilized mud, sand, and gravel lying above the epicenter, triggering a massive underwater landslide. Sediments in an area roughly 7,720 square miles, larger than the state of New Jersey, took off down the continental slope as a dense, turbid slurry. The underwater avalanche picked up more sediment and increased its speed as it flowed downslope. This turbidity current reached estimated speeds of 35–60 mph and deposited a sediment fan hundreds of miles from its source.

How do we know how fast the currents were moving? As the submarine landslide moved down the continental slope, it snapped twelve trans-Atlantic telephone cables. At the time of the 1929 quake, there were many submarine cables just off the coast of Newfoundland, where they crossed the increasingly deep water of the **continental shelf**, the **continental slope**, and the deep-sea basin. Telegraph companies recorded that the first six cables snapped at the time of the quake, while the next six, in increasingly deep water, broke sequentially over the next 13 hours. The companies recorded the exact time that each cable broke because communications stopped abruptly.

In 1952, two geologists at Columbia University's Lamont Geological Observatory used data from the 1929 quake to prove the existence of turbidity currents. Maurice Ewing and Bruce Heezen took a research ship into the area; they ran **seismic surveys** and collected **sediment cores** in submarine canyons off Newfoundland. They also accessed records from the telegraph companies that provided the timing of the cable breaks after the 1929 quake. Heezen and Ewing argued that a fault rupture would have broken all the cables at the same time. They concluded that only a turbidity current could have caused the sequential breaking of the trans-Atlantic cables.

PLATE TECTONICS MEANS CONSTANT CHANGE

The homeland of the Atsugewi runs along the Pit River east of Mount Shasta. In their creation story, Kwaw and Ma'Kat'da fought with each other for control of the mist, the material they used to create the world. Kwaw created; Ma'Kat'da destroyed, and in that creative maelstrom they made the landscape that came to be California. In much the same way, over the past 200 million years, the Pacific, North American, and Farallon tectonic plates collided and jostled to create the region's mountains and valleys, and gave rise to its climatic, topographic, and geologic diversity. Mountains captured rain and served as barriers to migration. The broken and diverse landscape allowed Indigenous peoples to develop communities with distinct lifestyles.

These original Californians had been here a long time: at Tulare Lake, in the southern San Joaquin Valley, archaeologists have uncovered Clovis points—thin, fluted spear points up to eight inches long that date to about 11,000 years ago. California's geology—the north–south Coast Ranges and Sierra Nevada and deserts—isolated these original settlers from cultures that developed in Mexico and the western United States. The broken topography made it difficult for Indigenous groups to travel great distances, and many groups tended to live in smaller family clans, unlike the larger tribes and nations to the east and south. California had the most diverse cultural groups, speaking as many as 135 distinct languages, of any similar area between the Arctic and the tip of South America. Varied geology meant varied topography and plant life, which influenced housing types: redwood plank houses along the Pacific Coast; conical homes of incense-cedar in the western Sierra; saltbush huts on the eastern slope of the Sierra; tule mat–covered homes along the San Joaquin delta waterways; and palm-frond houses around springs in the Mojave Desert.[1] Plate tectonics created the landscape that nurtured these diverse early California lifestyles.

Plate tectonics, the science of the Earth's **crust** and how it deforms, has caused a revolution in geologic thinking over the past sixty years. For undergraduate geology students in the late 1960s and early 1970s, plate tectonics was considered fringe science, and few American geologists bought into it. We were taught that mountains formed along the margins of continents, and that in this way continents grew outward and larger over time. And yet by the mid-1970s almost all geologists had been convinced of the correctness of the revolutionary new concept. The theory of plate tectonics states that the Earth's crust is made up of brittle plates, much like the cracked shell of a giant hard-boiled egg—except that these large plates jostle each other as they move slowly over the **mantle**, a zone of hot, viscous, and ever-moving rock below the crust. Roughly 250 million years ago, there was only one continent on Earth, a supercontinent called **Pangea**. Then upwelling mantle rock in the center of this supercontinent began to split the continent apart—the beginning of the Atlantic Ocean. West of the new Atlantic spreading center, the North American Plate began to move slowly westward.

What does that have to do with California? Just prior to Jurassic time, the western edge of what is now North America was a relatively quiet place, what in the field of plate tectonics is referred to as a **passive margin** (as opposed to an actively deforming margin like the present-day West Coast . . . think earthquakes). The margin was a gentle slope leading from the edge of the continent down to the deep ocean to the west. This continental shelf and slope accumulated great thicknesses of sediments eroded off the continent. (Most of those sediments are now found in Utah, Nevada, and Arizona, and a few remnants also exist in eastern California around the Sierra Nevada and Death Valley.) But the peace was not to last. Around 230 million years ago, a string of volcanoes, what's referred to as a **volcanic arc**, developed in the Pacific just off the west coast. Plate tectonics explains why.

An oceanic spreading center opened in the western Pacific and began creating new oceanic crust. As new crust formed at the mid-ocean ridge, the older crust moved away from the ridge. Oceanic crust west of the spreading center, the **Pacific Plate**, began moving west, and crust east of the spreading center, the **Farallon Plate**, moved slowly east. **Basalt**, a black lava that is the main component of oceanic crust, is rich in iron minerals, making it more dense than **continental crust**, which is mainly granite containing lighter aluminum and silica minerals. When the dense oceanic crust arrived at the western margin of the continent, it slid under the North American Plate. Or the continent, being lighter, overrode the oceanic crust. It all depends on your point of view.

We all know, in an abstract way, that the Earth gets hotter the deeper you go into the subsurface. The actual thermal gradient is about 29°F/mile, which may not sound like a lot, but if you were to go into the deepest mines, the heat would quickly become unbearable. The deepest mine on Earth, the Tau Tona gold mine in South Africa, goes down over 2 miles, and the rock is 140°F. Despite excellent ventilation, miners can only work for a couple of hours. The deepest well on Earth, the Kola Superdeep Borehole in Russia, got down to 40,200 feet, where the temperature was a whopping 356°F. At those temperatures, rocks get soft, similar to the way rock-hard caramel candy out of the refrigerator soon becomes soft and malleable when you pop it in your mouth.

Add pressure to the softened rocks, and their structure begins to change. As Morgan Freeman's character in *The Shawshank Redemption* says, "Geology is the study of pressure and time. That's all it takes, really. Pressure... and time."

One way to measure pressure is in pounds per square inch (psi). The pressure in the Earth, due to the weight of overlying rocks, increases at an average of 1 psi for each foot beneath the surface. This is just like water pressure increasing on your ears when you dive to the bottom of a swimming pool: it doesn't take long before it's painful. At 20,000 feet (about four miles, the depth of many oil wells), the pressure is 10 *tons*/square inch pushing down on the rock. The temperature at that depth is already around 165°F. (Just for perspective, the hottest temperature ever recorded at the Earth's surface was 134°F at Furnace Creek, Death Valley. The hottest I've ever been in is 115°, and I can affirm that it was *not* comfortable.)

The Earth's crust is thicker on continents (18–30 miles) and thinner in oceans (3–6 miles). The contact between the base of the crust and the upper mantle is both hot and under a lot of pressure: temperatures range from 400°F to 750°F, and pressures run from 8 tons/square inch to 80 tons/square inch.

The Earth's mantle is hot and quite dense, and the cooler and lighter crust floats on it, like ice cubes on water. The heavy and dense oceanic crust floats low in the mantle and is covered by the oceans. The lighter, more buoyant continents float higher on the Earth's mantle, and only their margins are beneath the sea. I like to think of the continents as the frothy scum of the Earth.

The internal heat in the Earth causes forces that push up and sideways. The upward and sideways movement is part of a **convection current**, and it drives plate tectonics. An ocean spreading center works like the convection in a boiling pot of water: the hot water rises and pushes outward, then cools and sinks down along the edge of the pot. At a spreading center, hot rock rising from the mantle comes to the surface and spreads outward; after it has moved away from the mid-ocean ridge and cools, it sinks back into the mantle.

The process of an oceanic plate plunging under a continental plate is called **subduction**. As an oceanic plate sinks into the Earth's mantle, it forms a dip in the surface that is an **oceanic trench**. These are the deepest parts of the ocean floor. The area of the trench and down-going ocean floor is called a **subduction zone**. These trenches accumulate sediments eroded off the continent. Some of these water-logged sediments are dragged down into the mantle along with the descending plate, and melt. Most of the sediments, along with slivers of the ocean floor, get crumpled up on the edge of the continent. Think of it as trying to slide a pizza under a door: most of the topping is going to get scraped off and pasted onto the door. That topping is the **Franciscan Terrane** (also referred to as Franciscan Complex, Franciscan Mélange, or Franciscan Assemblage) that makes up much of western California.

The Farallon Plate melted as it went down into the mantle, and the resulting magma was buoyant enough to rise up through the **continental margin** and form a volcanic mountain range. At that time, the continental margin would have appeared much as the Aleutian Islands appear today, as an arc of very high, very active volcanoes. The deep roots of the old volcanoes are the granitic rocks of the Sierra Nevada, the sawtooth mountain range near the California-Nevada border. The old passive-margin sediments that the magma passed up through were altered by heat and pressure. These altered sediments and the volcanoes above the granites are now mostly eroded away, but a fringe has been preserved in the western foothills of the Sierra, and I have seen local bits above the granite on some of the higher peaks. They look like giant chunks of rock, ranging from red-brown to black, floating above or in a mass of white granite. It's as if you are in an old crumbling cabin where bits of the roof have collapsed into the ceiling above you. The roof rocks, or **roof pendants**, are the old continental shelf sediments. The Sierra volcanic arc is where the edge of the continent used to be.

When I refer to granite in the Sierra Nevada, it isn't the granite that I have on my countertops at home. What homebuilders call "granite" can be any rock, from **marble** to granite to man-made material. Sierra Nevada granite, to a geologist, is technically a **granodiorite**, intermediate between a true granite and a **diorite**. True granite, to those familiar with Pikes Peak in Colorado or the Llano Uplift in Texas, tends

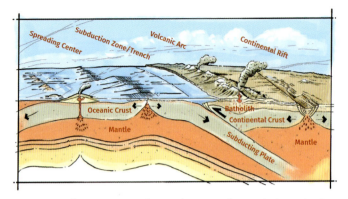

Tectonic plates are created at mid-ocean ridges and plunge back into the mantle at subduction zones. San Francisco would have been perhaps 100 miles west of the volcanic arc (Sierra Nevada) some 25–30 million years ago when subduction ended.

to be orange-pink in color, a result of potassium-rich **feldspar**. A granodiorite, in contrast, lacks the pink feldspar and only has white feldspar, making it speckled black-and-white, the so-called salt-and-pepper granite. Whenever I mention Sierra Nevada granite, I am really referring to the bedrock in general, which is mainly granodiorite.

West of the Sierra Nevada is California's Central Valley, filled with marine sediments called the **Great Valley Sequence**. Driving through the Central Valley is tedious, dull, mind-numbing. It's flat and hot and covered with miles and miles of fruit and nut orchards. The valley is flat and geologically boring because it used to be a shallow ocean that has since been filled to the brim with sediments. Rocks that eroded off the volcanic arc and Sierra Nevada form the bulk of the 65–150-million-year-old Great Valley Sequence. The valley started as a deep marine basin, but got shallower as it filled. The seashell-rich marine mudstone and sandstone that make up the Great Valley Sequence is a bit over 8 miles thick. Eight miles is the upper range of cruising altitude for most commercial aircraft—in other words, the valley is filled with a lot of dirt.

A Very Brief Introduction to California Geology 13

This diagram is our best guess at what western North America looked like 145 million years ago, during Late Jurassic time. The map shows about where San Francisco is today. Please note that the line below the words "Next Figure" indicates the location of the cross section shown in the figure following this one.

West of the Great Valley Sequence is the Franciscan Terrane. It formed way out in the Pacific, far off the west coast of North America. In most parts of the world, rocks form nice, continuous layers, with the oldest layers on the bottom and the youngest on top. The Franciscan is not the typical layered geologic formation. Pillow lavas, siliceous ooze, and turbidites were all deposited in the Jurassic sea. These form what I call the **Franciscan trilogy**, a discontinuous mash-up of rocks that were scraped off the ocean floor and accreted, or pasted onto the continent during the process of subduction. This dog's breakfast of rocks is what geologists call **mélange**, from the French word for a mixture of bits and bobs.

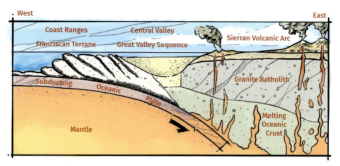

Subduction and creation of the Franciscan Terrane prior to about 30 million years ago

Subduction along the western margin of North America began about 230 million years ago, in the Triassic period, and lasted about 200 million years. During this huge interval of time, as noted earlier, ocean sediments slowly accumulated on and were carried by the Farallon Plate into an offshore trench, where they were welded onto the western edge of North America.

The process of oceanic spreading, subduction, and volcanic arc formation is still going on both north and south of California. Ever wonder why there are active volcanoes in Washington, Oregon, and northernmost California, but no volcanism in most of California, and then the volcanoes start up again in Mexico and continue through Central America and the Andes? The "Ring of Fire"—the volcanoes that surround most of the Pacific Ocean—exists because oceanic plates are still going down and melting under most of North and South America (and in the western Pacific). But sometime between about 25 and 30 million years ago, something changed in California. The west-moving continent began to override not just the oceanic plate, but the actual spreading center, the **East Pacific Rise**. Wherever the continent rode over the spreading center, this action put an end to subduction and volcanism. Instead of subduction, a different kind of margin formed between the continent and the Pacific Plate. The margin became a **shear zone** or **transform fault**,

A Very Brief Introduction to California Geology 15

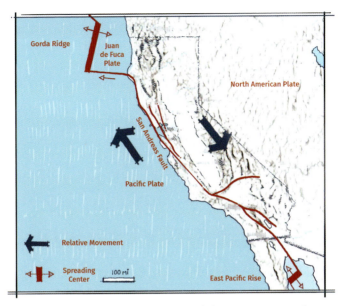

The San Andreas Fault system extends between two spreading centers, the East Pacific Rise and the Gorda–Juan de Fuca Ridges. The East Pacific Rise terminates at the north end of the Gulf of California. The Gorda Ridge lies off the shore of Northern California; the Juan de Fuca Ridge lies off the coast of Oregon and Washington. All three of these rises are remnants of the Pacific spreading center.

the San Andreas Fault system, where the two sides of the fault slide sideways past each other. You can visualize this if you put your hands side by side flat on a tabletop. Now slide your left hand up and your right hand down: this is the San Andreas Fault in California. The side closer to the ocean moves northwest, and the continent side moves southeast. But it's not just one monster fault. Rather, it is a wide zone of faults that includes the infamous San Andreas Fault, the far less famous (but arguably more dangerous) Hayward Fault, and many others. Some faults related to this zone have been mapped as far east as Nevada. West of each related fault, the rocks are moving slowly northwest, while east of each fault, the rocks are moving southeast. Generally speaking, fault movement in this zone is older in the west and younger to the east, meaning that this broad shear zone is migrating eastward over time. This is the situation that affects the west coast of North America from the Gulf of California to the Mendocino coast.

So, contrary to what we have all heard many times, California is not going to fall into the sea. Rather, in a few tens of millions of years, those of us living along the California coast will have to trade in our surfboards for snowmobiles as we approach Alaska.

Eager to start my explorations, I grab my hat, a camera, and a notebook, and head out to see the land.

CHAPTER 1

Coast Ranges

Point Bonita and the Marin Headlands, just north of San Francisco and across the Golden Gate, highlight two of the rock types that make up the California Coast Ranges. The Point Bonita Lighthouse parking area, where we begin our journey, is located at the end of Conzelman Road at latitude 37.821905 and longitude −122.529489.

MARIN HEADLANDS

Marin Headlands, jutting into the wild Pacific, is known for its outstanding natural beauty as well as its unique geology. For this reason, it is included in the Golden Gate National Recreation Area. The bluffs and sea cliffs are part of the Franciscan Assemblage, the mishmash of different rock types that formed far out in the Pacific over the past 200 million years. Walking across the low hills, green in winter, clothed in flowers in spring and in dry yellow grass the rest of the year, I find all of the main components of the ocean floor: the Franciscan trilogy of pillow lavas, ribbon chert, and graywacke sandstone. The pillow basalts formed along the Pacific spreading center, where ocean floor was created. In the time since the pillows formed, the originally black basalt lava has changed, by exposure to seawater and hot fluids in the Earth's crust, so that it now appears greenish. Minerals that form at great depth are only stable at high temperatures and pressures. When they end up at the Earth's surface, they are highly unstable, and weathering can change their chemistry and crystal structure to more stable minerals. Thus the green colors are a result of some of the original minerals (dark **pyroxene**, **olivine**, and **amphibole**) altering to the green minerals **actinolite** and **serpentine** and the clay minerals **chlorite** and **glauconite**. Greenish pillow basalt underlies the Point Bonita Lighthouse.

Above the dark green-gray basalt are rusty-brown radiolarian ribbon cherts, relics of Jurassic to Cretaceous seas, roughly 200 to 100 million years ago. These are some of my favorite local rocks because they reveal, in their intricate **folds** and swirls, the deformation that these rocks have endured. Ribbon chert is exposed along Conzelman Road.

Above the chert is the youngest rock, graywacke sandstone. Excellent **outcrops** are where fresh rock, unweathered and unspoiled by plant or soil cover, is exposed for all to see. Such outcrops can be seen at the north end of Rodeo Beach. Taken together, this rock trilogy of pillow lava, ribbon chert, and graywacke sandstone tells a story of evolving marine environments, from a deep tropical ocean to a temperate continental margin.

The topic of **metamorphism** and **metamorphic rocks** will come up over and over as we traverse the state. Metamorphism is any process that changes rocks and their constituent minerals from one form to another as a result of changing pressure, changing heat, and/or hot, mineral-rich fluids moving through the rock. Under intense pressure, the mineral grains in a rock begin to align. This layered (foliated) rock fabric indicates the degree and direction of pressure on the rock and is one of the key indicators of a metamorphic rock. The altered, metamorphosed rocks are called **metasediments** or **metavolcanics**, depending on what the original rocks were.

Start: Point Bonita Lighthouse parking area, Marin Headlands (37.821905, −122.529489).

STOP 1: PILLOW BASALTS AT POINT BONITA LIGHTHOUSE

Walk about 1,300 feet (a quarter mile) down the trail to the lighthouse. We are walking by rocks formed at a mid-ocean ridge beneath miles of water. Point Bonita is made of dark-green Franciscan pillow basalts. These Jurassic volcanic rocks erupted at the Pacific spreading center. They have since been metamorphosed—that is, altered to rocks rich in green minerals. And there they are: green pillows of altered basalt weathering out of the hillside beside the trail.

Point Bonita Lighthouse is still active and is maintained by the US Coast Guard. It was the third lighthouse built on the West Coast: the others were at Fort Point and on Alcatraz Island. Apparently, the local fog caused many of the old ships to founder on the rocky headlands: over three hundred ships ran aground during the Gold Rush years alone. Built in 1855, the lighthouse was originally located 188 feet higher on a hill above here, but there were too many complaints that the beam was lost in the fog. The original tower was moved to the

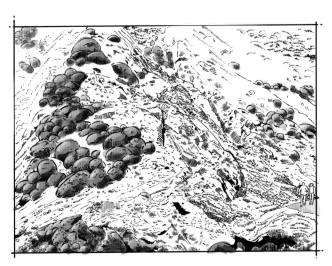

Pillow lavas (the blobs ranging from dark gray to greenish) along the trail to the Point Bonita Lighthouse

exposed point in 1877. A tunnel 118 feet long had to be hand chiseled through the hard rock, after which a wooden walkway to the lighthouse was built. When the cliffs around the walkway began to fall away, a suspension bridge was installed in 1954. By 2010, salt spray had rusted out the original bridge, and it had to be replaced. The new bridge, completed in 2012, is made of tropical hardwood planks and steel cables.

In dense fog, when the light signal can't be seen, a foghorn supplements the light. Originally a cannon was used, but it proved ineffective and was replaced by a fog bell around 1859. Later, about 1874, a steam siren was installed. Today, the lighthouse has an electric foghorn. The foghorn has a distinct signal, sending out two blasts every 30 seconds; it is triggered by a laser fog detector.

Visit the website for open hours and the schedule for ranger-led hikes. Access is free: https://www.nps.gov/goga/pobo.htm.

While I am in the area, I inadvertently bump into some intriguing Cold War history hiding in plain sight. Nike Missile Site SF-88L is just off Field Road, less than a mile north of the Point Bonita Lighthouse parking lot. If you are in the area, try to catch the tour. This is a decommissioned missile battery with several of the 1960s-era antiaircraft missiles. During the tour, the docent explains how they were meant to be used. Apparently, radar guidance in the time of vacuum tubes was not good enough to intercept Soviet bombers directly, so the plan was to launch the missiles while the enemy bombers were still 100 miles off the coast, then detonate nuclear warheads to wipe them out.

It seems obvious that this plan did not take into account the prevailing westerly winds and what would happen to any radioactive fallout that would drift over San Francisco and the Bay Area that they were trying to protect. Maybe they thought a little fallout was better than the alternative.

Visit the website for tour hours. The tour is free: https://www.nps.gov/goga/nike-missile-site.htm.

From the Point Bonita Lighthouse parking area, drive 3.3 miles (about 10 min) east to the Coastal Trail parking area at the junction of McCullough Road and Conzelman Road (37.833826, −122.494396). Walk south on Conzelman Road about 1,250 feet to the ribbon chert roadcut.

STOP 2: DEFORMED FRANCISCAN RIBBON CHERTS

When I drive by great roadcuts, I have a tendency to stare at the rocks instead of the road. It's a weakness that I have struggled to overcome, to no avail. I admit it: while glancing at an outcrop, I tend to let the car wander across the centerline or crunch onto the shoulder of the road. It is at this point that my wife will say, in gentle but firm tones, "You look at rocks; I'll drive."

At Conzelman Road, I find the second in the trilogy of Franciscan rocks: the ribbon chert. The sudden change from pillow basalt to ribbon chert in this short distance highlights one of the defining attributes of a mélange: the abrupt and discontinuous nature of these rocks.

This spectacular outcrop exposes deformed ribbon cherts. I particularly like the folding that is highlighted by the thin layers. The rock looks as though it's been squeezed out of a giant tube of toothpaste. Which it kind of has been, because the rocks were buried so deep and under such high pressures that they became soft and pliable and readily folded.

These are some of the best exposures of ribbon chert seen anywhere. The layers, composed of countless silica shells of radiolarian plankton, were welded together by the pressure of miles of overlying sediments and seawater. The silica ooze became chert, a glassy form of quartz. After being buried and squeezed in the subduction zone, these sediments were shaved off the top of the down-going oceanic plate, like a block plane shaving a plank of wood. The rock was then pushed up and to the west by the North American Plate. The entire sequence was piled up like snow being shoveled off a driveway. This was accomplished by thrust faulting.

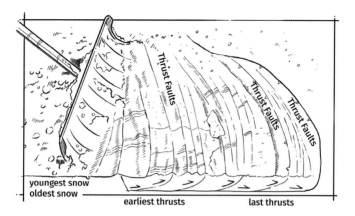

Snow-shovel analogy for illustrating the formation of a thrust-faulted mélange terrane. In this case the snow piles up in front of the shovel. In the trench, slivers of ocean floor and overlying sediment pile up as the continental plate plows westward.

Most people think of faults as vertical breaks in rock that drop one side down and elevate the other side. That kind of fault forms fault-bounded mountain ranges or valleys, and they are called **normal faults**. Another type of vertical break, where the two sides move horizontally past each other, is like the San Andreas Fault. Geologists call these **strike-slip faults**. A third type of fault pushes older rocks upward over younger rocks. These tend to be nearly horizontal breaks, and are called **thrust faults**, because one side is thrust over the other side. These are the faults that caused the deformation within the Franciscan Terrane: they brought sediments from the depths of the subduction zone all the way to the surface, all the while breaking them into small, coherent packets of rock and folding them. The folded chert at Conzelman Road reveals and embodies the internal deformation in this pile of marine sediments.

Coast Ranges 23

Deformed ribbon cherts along Conzelman Road

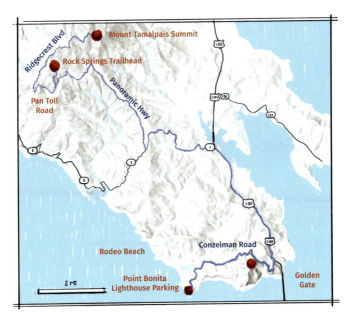

Map of the route from Coastal Trail to Rock Springs and Mount Tamalpais Summit

MOUNT TAMALPAIS AND SAN FRANCISCO BAY

Mount Tamalpais (from the Miwok word for "west mountain" or "bay mountain") is a major Bay Area landmark with a profile that is easily recognized from the East Bay. The 2,579-foot peak is the highest point in Marin County and provides panoramic views of San Pablo and San Francisco Bays. Mount Tam's distinctive profile is a result of gentle south-facing slopes and steeper north slopes. The steep flank of the mountain, combined with almost 53 inches of annual rainfall, causes numerous landslides. Old landslides reveal themselves by their hummocky terrain and the numerous ponds found in the sags.

Early settlers, their imaginations perhaps charged by a lack of female companionship, came up with a legend to explain Mount Tam's profile. They told of a beautiful young maiden who was in love with an Indian prince. When he abandoned her, she walked to the top of the nearby mountain, where she died of heartbreak. The mountain was so moved that it

Mt. Tamalpais from Ring Mountain (*looking south*)
(Woodcut © Tom Killion. Reproduced with permission.)

changed its form, taking on the outline of her body. Ever since, Mount Tam has been referred to as the Sleeping Lady. It should be noted, however, that this name is not rooted in any known Native creation story.[1]

The mountain is made up of uplifted and deformed Franciscan Assemblage rocks that include basalt, chert, graywacke, and **shale**. Roadcuts expose the tightly folded red-brown chert. In some places, the ribbon chert has been altered, bleached white by ancient hot springs. In other places, it has been turned into red jasper. But we are here for another reason. At Mount Tam, we go deep. Literally. I want to touch the guts of the Earth, actual mantle rock.

From Marin Headlands, proceed to Mount Tamalpais to find outcrops of serpentinite and come upon expansive views of San Francisco Bay. Drive to the Rock Springs Trailhead (37.910663, −122.612667), 15.3 miles (about 40 min) from the Conzelman Road stop. Park in the small lot west of Pan Toll Road and stroll about 725 feet southwest to an outcrop of serpentinite. You'll recognize it by the green color and broken appearance. The rock has slick, polished surfaces and a soapy or waxy feel.

I came of age at a time when paper maps were the only thing that kept me from getting hopelessly lost. These maps were clumsy, and hard to hold and look at while driving: you either needed a navigator or had to pull over to peruse them. I'd mark my route with yellow highlighter and circle important turnoffs. Folding the maps so that I saw only the area of interest was a form of origami. I have to say I have liked the advent of Google Maps, Apple Maps, and onboard global positioning system (GPS) navigation over the past decade. The maps are easy to glance at, and you can get them to talk to you, telling you how far till the next turn and which way to turn. Some will tell you the speed limit, which is not always apparent from road signs. Yes, there are times when I have to ignore the instructions because a road has changed or there is new roadwork, but that is fairly uncommon. Before heading to the next stop, I just plug in the latitude/longitude or street address, and the car's GPS unit or my phone takes me right there.

STOP 3: ROCK SPRINGS TRAILHEAD SERPENTINITE
The road up Mount Tamalpais follows the curves of the mountain. The narrow, steep road snakes its way through damp pine forests and dark redwood groves. The sunlight filters down to the road like a fine mist. On the drive up, I pass bikers straining against the mountain, as if pushing harder will make it give way. It doesn't.

It may be geological sacrilege to say, but there is more to rocks than rocks. It is reassuring to know that rocks support, are a foundation for, and provide nourishment to the plants that grow above them. So it is with the giant redwoods that require mineral soil to germinate their seeds. These trees are throwbacks to the Pleistocene when everything was supersized. The coast redwood, *Sequoia sempervirens*, is the tallest living thing. At 300 feet plus, it is almost impossible to take it all in; you have to lean back to see the top. The tallest of these trees is 60 feet taller than the Statue of Liberty. These russet monsters have been corralled by glaciers and felled by loggers, and they've had their range diminished until there are almost no old-growth forests left. In 1966 Ronald Reagan was running for governor of California and embodied the timber industry's feelings at the time:

> I think, too, that we've got to recognize that where the preservation of a natural resource like the redwoods is concerned, that there is a common sense limit. I mean, if you've looked at a hundred thousand acres or so of trees—you know, a tree is a tree, how many more do you need to look at?[2]

The balancing of commercial interests against resource preservation *is* a legitimate issue, but the blunt expression of Reagan's thoughts came across as "If you've seen one redwood, you've seen them all." Luckily, a few large stands were preserved in Redwood National and State Parks and Muir Woods National Monument, among others.

Their mountain cousins, *Sequoiadendron giganteum*, the giant sequoia, were also logged, mostly in the 1800s, but it turned out that the wood was too brittle, and most of the trees splintered when they fell. It is said the wood was good for little other than grapevine stakes, a sad end for such a colossus. *Giganteum* are not the tallest trees, but do have the greatest girth, and are among the oldest living things, some pushing 3,300 years. It is hard to photograph a person standing next to one of these trees, as they are dwarfed by the tree. Today these ice age relics are confined to a few protected Sierra groves extending from Yosemite's Mariposa Grove to Giant Forest in Sequoia National Park.

Trees aside, it is worth coming to Mount Tam for the **serpentinite**. I literally stumbled onto this broken outcropping the first time I hiked the Rock Springs Trail up Mount Tamalpais. The rock is fractured, with polished surfaces that reveal its origins in a shear zone. It is markedly green, and plants don't grow well on it; apparently the minerals in the rock are objectionable to most of the local flora. Serpentinite is attractive and rare, a rock that ranges from apple green to black that also happens to be the official state rock of California. An exceptional outcrop of serpentinite occurs near the Rock Springs trailhead on the southwest flank of Mount Tamalpais.

Serpentinite is uncommon at the surface of the Earth. It starts out as a rock called **peridotite**, the commoner cousin of the yellow-green gemstone **peridot**. Peridotite forms deep in the Earth, 60 miles or so beneath the surface. Occasionally, as in subduction zones, mantle peridotite gets caught up in the collision between ocean and continent and works its way up to the surface. As it inches upward, groundwater penetrates the minerals, and these hydrated minerals convert the peridotite to serpentinite. The rock has a waxy, smooth, glossy appearance, the result of being squeezed and sheared on its way to the surface.

Serpentine was an important source of asbestos before people learned about the hazards associated with that mineral. But don't worry about touching or holding serpentinite. As one geologist put it, the only way a piece of serpentinite can hurt you is if someone throws it at you.

Serpentinite outcrop near the Rock Springs trailhead

Drive northeast on Ridgecrest Blvd 3.0 miles (10 min) to the summit of Mount Tamalpais (37.927246, −122.579997). Park at Mount Tamalpais East Peak parking lot for commanding views of San Francisco Bay.

STOP 4: MOUNT TAMALPAIS SUMMIT AND THE EVOLUTION OF SAN FRANCISCO BAY

A short trail leads to the classic old stone fire lookout atop the East Peak of Mount Tamalpais. The views, especially on a clear day, are sweeping and seemingly endless in all directions. Even when there is fog on the Bay, the views are magnificent: it is usually possible to see the tops of the towers in San Francisco and the upper parts of the Golden Gate Bridge above a snowy-white carpet of low clouds.

When I look out over the Bay, it is hard not to think it has always been here. And yet, as I learned, San Francisco Bay is a recent affair, geologically speaking. A million years ago, the area that is now San Francisco Bay was a flat plain between the Coast Range and the **East Bay Hills**. Half a million years ago, rivers draining the Central Valley had established a path across the plain to the Pacific. But not through the Golden Gate. The rivers cut through the Coast Ranges at Colma Gap, just north of San Francisco International Airport.

The Coast Range, the backbone of the San Francisco Peninsula, formed as buckles along the San Andreas Fault. As the fault moved sideways, the rocks on either side crumpled, kind of like trying to slide two sides of a carpet sideways past each other. As it turns out, about 125,000 years ago, the San Andreas Fault slipped sideways, and a ridge of land dammed the river flowing through the Colma Gap. Eventually the river found and exploited another low spot, a new gap in the Coast Range just north of San Francisco. The Golden Gate.

The entrance to San Francisco Bay was not always called the Golden Gate by settlers. Juan de Ayala and the crew of the *San Carlos*, the first Europeans to sail into the Bay in 1775, named it Boca del Puerto de San Francisco (Mouth of the Port of San Francisco). I had always been under the impression that the name was changed during the Gold Rush, because it was the gateway to the gold fields. Turns out that in 1846, two years before gold was even discovered in California, US Army Captain John Frémont was in Mexican California on an official US government survey of the West. When he saw the entrance to the Bay, he commented that "it is a golden gate to trade with the Orient." The first published use of the name was in his *Geographical Memoir*, which he submitted to the US Senate in 1848. Visualizing rich cargoes from the Orient, he wrote: "to this Gate I gave the name of 'Chrysopylae,' or Golden Gate, for the same reasons that the harbor of Byzantium was called Chrysoceras, or Golden Horn."[3]

But I digress. During the last interglacial period, about 114,000–125,000 years ago, sea level was actually higher due to the lack of ice sequestered in **ice sheets**. San Francisco Bay was even larger than it is today. We know this because there are scattered shoreline deposits and remnants of a **marine terrace** about 20 feet above the current water level. A marine terrace is the flat surface carved by wave action against a shoreline. When left high and dry by a falling sea level, they are recognized as flat surfaces mantled by mud and sand. Downtown Oakland is built on one such natural platform.

At the height of the last ice age 20,000 years ago, so much seawater was tied up in ice that sea level was 400 feet lower than it is today. The coast was 20–30 miles west of where it is now, and the Farallon Islands were mountains along the coast. The river that drained the Central Valley had cut a deep canyon through the Carquinez Strait in the East Bay Hills to the Golden Gate in the Coast Range and westward to the coast. Where most of the Bay is today was a broad river valley 15 miles wide and 300–400 feet deep.

Between 12,000 and 15,000 years ago, Earth's climate started to warm and the glaciers began to melt. Sea levels rose, sometimes as much as an inch a year. As sea level rose, sediment carried by the combined San Joaquin–Sacramento River was deposited in Golden Gate Canyon, backfilling it. Eventually, about 9,000–10,000 years ago, the rising Pacific spilled through the gap at the Golden Gate and began to fill the valley between the Coast Range and the East Bay Hills. This was the beginning of San Francisco Bay. Perhaps not surprisingly, the average depth of the Bay is only 12–15 feet, depending on tides. The deepest part of the Bay is under the Golden Gate Bridge, where twice-daily tides scour the old valley sediments to a depth of 372 feet.

Sea level is still rising, but sediments from the surrounding hills are trying to fill the Bay. Several hundred feet of sediment

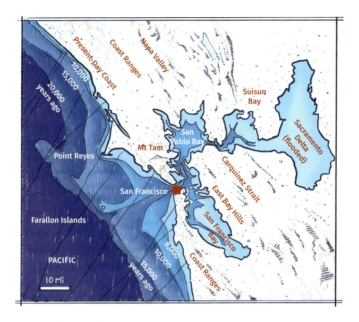

Evolution of San Francisco Bay over the past 20,000 years

have been deposited in the Bay in just the past 10,000 years. Deposits from numerous rivers and creeks around the Bay accumulated in deltas along the shore, forming extensive mudflats, tidal marshes, and occasional sand dunes.

The original Ohlone inhabitants lived a relatively peaceful existence along these shores, hunting deer, gathering acorns, fishing, and harvesting shellfish. This was the situation when the Spaniard Juan Gaspar de Portolá and his overland expedition, on a mission to explore, colonize, and convert Indigenous people to Catholicism, reached San Francisco Bay in November 1769.

Since that time, the Bay's shoreline has been significantly modified by sediment washed down from the goldfields by **hydraulic mining** in the mid-1800s. In 1917, G. K. Gilbert of the US Geological Survey estimated that over a billion cubic yards of mining debris had washed into the estuary from Suisun Bay to San Francisco Bay.[4] More recently, artificial landfills destroyed the shallow coastal marshes in favor of land

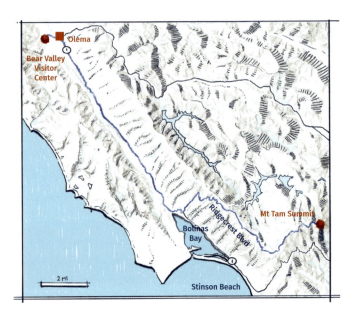

Map of the route from Mount Tamalpais to the Bear Valley Visitor Center

for human structures. There would be no San Francisco Bay but for the Coast Range, and there would be no Coast Range were it not for the San Andreas Fault. Wherever the two sides of the fault push against each other, mountains are raised up.

THE SAN ANDREAS FAULT

For a geologist, it doesn't get much better than this. There are not many places where we can walk on the edge of a tectonic plate, but a few miles north and west of Mount Tamalpais, I am able to.

The North American Plate does not coincide with North America. Rather, it is presently forming at the Mid-Atlantic Ridge spreading center and includes all of the western Atlantic as well as North America. Many geologists make the pilgrimage to the eastern margin of the North American Plate, a long, narrow **rift** valley at Thingvellir, Iceland, where the North American Plate is separating from the European Plate. It is a whole lot easier to walk along the western margin of the North American Plate: the San Andreas Fault at the Bear Valley Visitor Center in Olema.

From Mount Tamalpais, drive northwest to the town of Olema and turn west onto Bear Valley Road. Drive to Bear Valley Visitor Center Access Road on the left and park (38.041394, −122.799860), for a total of 21.4 miles (50 min).

STOP 5: BEAR VALLEY VISITOR CENTER AND THE SAN ANDREAS FAULT

I want to stand atop the mighty San Andreas Fault. This is a fault like few others in the world, the very embodiment of a transform plate margin. East of the fault is Franciscan Terrane. West of the fault is Salinian granite and sediments that originated far south of here. I want to put one foot on the Pacific Plate, the other on the North American Plate. I can do that here, on the Earthquake Trail.

The visitor center is in a low spot in the valley, a clearing in the coastal forest where colorful California quail chitter and skitter among the picnic tables. The low valley marks the trace of the San Andreas Fault zone, almost a mile wide here. The ground is quiet most of the time, but there is tension in the air. At any moment, without warning, the land could rip apart, jumping up and sliding sideways at the same time.

A few hundred yards north of here, Tomales Bay cuts a long, linear swath across the countryside. Tomales Bay was created by sideways movement along the San Andreas Fault. Though the two sides of the fault mostly slipped parallel to each other, in this area there was just a bit of separation that caused the bay to sag downward. The linear trace of the San Andreas Fault is especially clear when observed from an airplane. Long linear landscape features such as **scarps**, sag ponds, and Tomales Bay reveal the faulting here.

In 1906, at the same time that a massive quake along the San Andreas Fault effectively destroyed San Francisco, the ground here broke and jumped sideways. This really was "the big one." The quake was centered just offshore from San Francisco, 25 miles south of here. There were fewer people and homes in this rural area, but the effects of the magnitude 7.7–7.9 quake were still impressive. In the blink of an eye, the ground ruptured and the Pacific Plate moved sideways as much as 20 feet. The North American Plate lurched upward a couple of feet. Fences were offset. People were all shook up.

Do not be deceived by the peace and quiet at this stop. It is not logical to think that a process that has been going on for the past 25 million years would suddenly end. Movement on the San Andreas Fault *will* happen again. The only question is when.

The effects of fault movement over time are written across the landscape. In addition to creating Tomales Bay, the fault caused the granite lying under Inverness Ridge to rise to the surface. One of the eminent early American geologists, Grove Karl Gilbert, was asked by President Teddy Roosevelt to study the Great San Francisco Earthquake and submit a report. One of his stops was Olema, here at the Earthquake Trail. Of the fault's local effects, Gilbert wrote that there was "an extensive shifting of mud on the bottom of Tomales Bay.... The horizontal change of position near the southwest shore was in places more than 25 feet, and the vertical change as much as 2 feet" and that "the only notable water waves generated by the shock were in Tomales Bay, where a group of waves estimated to be 6 or 8 feet high came to the northeast shore."[5]

It is worth taking a moment to examine what an earthquake actually is and why there are so many in California. Active tectonic zones, mostly near plate margins, are the most likely places for earthquakes, and California contains a plate margin along the San Andreas Fault that slices across the state from northwest to southeast.

Tectonic plates, being at the Earth's surface, tend to be cool and under low pressure, so they undergo mostly brittle deformation when under stress. Brittle deformation means that the plates are inclined to rupture, or break along faults. An earthquake is the result of movement along a fault that releases pent-up stress. The release of stress causes sudden, sometimes violent shaking. The vibrations caused by that break spread outward as waves, much as ripples spread outward when a pebble is tossed into a pond.

Earthquakes happen all the time everywhere on Earth. Most are so small that we don't feel them. There are far more small earthquakes than large ones. Earthquake size, the amount of energy released, is measured using the **Moment magnitude scale**, which replaced the well-known Richter scale. The Moment magnitude scale is more reliable than the Richter scale for very large earthquakes, as it measures the size or energy of the earthquake at its source. As with the Richter scale, an increase of one step on this scale corresponds

Trace of the 1906 fault at Olema. Sideways offset here was 20 feet. The photo is from the G. K. Gilbert et al. 1907 report.

to a 32-fold increase in the amount of energy released, and an increase of two steps corresponds to a 1,000-fold increase in energy. Tremors less than 3 on the magnitude scale are rarely felt and are usually nondestructive, but they can be recorded. Values greater than 4 start damaging structures. The largest earthquake ever recorded was a magnitude 9.5 monster off the coast of southern Chile in 1960 that caused **tsunamis** felt as far away as Japan.

Getting back to the San Andreas Fault, this plate boundary is characterized by frequent small shifts and rare large displacements. Over time, these movements accumulate and cause major disruptions to the landscape. Point Reyes is one of these.

POINT REYES NATIONAL SEASHORE

Point Reyes is unique. Not just because of the windswept, dune-laced shore and scenic lighthouse. No, it's because the rocks are not from around here. In fact, the rocks are unlike anything within a hundred miles. Since crossing the San Andreas Fault back at the Bear Valley Visitor Center, I have been on the Pacific tectonic plate. The Pacific Plate has been inching its way slowly northwest in this area for maybe as long as 25 million years, since the end of subduction in this area. Over such vast periods of time, inches add up to miles. As Simon Winchester said in his book about the 1906 earthquake, "The beauty of California is inseparable from its seismic activity, as if the land itself is in constant motion."[6] Point Reyes is proof that it is.

To get to Point Reyes National Seashore, take Bear Valley Road to Sir Francis Drake Blvd and continue to the lighthouse parking area (37.997798, −123.012088), for a total of 20.6 mi (40 min).

STOP 6: POINT REYES LIGHTHOUSE

For me, Point Reyes is a favorite place to take guests. On a clear day, the views are stunning. Looking west from the

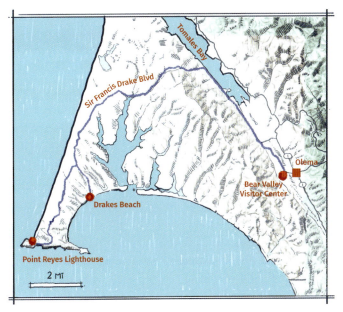

Map of the route from the Bear Valley Visitor Center to Point Reyes and Drakes Beach

lighthouse are the wind-tossed breakers that have moved across the Pacific from way beyond the horizon. If you are lucky, you might see whales breaching or blowing their misty breath into the air. To the north, white-sand beaches and dunes stretch 10 miles. I look for the herds of tule elk grazing on the low scrub and grass clinging to the sea cliffs. The story of these elk is one of grit and survival.

Tule elk are one of two subspecies of elk native to California. By the mid-1870s their numbers were reduced by hunting to fewer than thirty animals in a single herd near Bakersfield. A conservation-minded cattle rancher, Henry Miller, decided to preserve this last group. In 1978, the California Department of Fish and Game, along with the US Fish and Wildlife Service and the National Park Service, reintroduced eight female and two male elk to Tomales Point at the far north end of the Point Reyes peninsula. Another herd was established at Point Reyes in 1998 using twenty-eight animals from the Tomales Point herd. The elk appear to thrive in this environment: the Point Reyes group averaged 420 head between 1999 and 2019.

The frequently fog-shrouded Point Reyes headland extends 11 miles into the Pacific. This posed a threat to ships traveling north from San Francisco during the 1800s. Point Reyes Lighthouse was built in 1870 to warn ships away from this navigation hazard. In 1975, the US Coast Guard installed an automated light next to the historic tower and retired the original lighthouse. At that time, the Coast Guard transferred custody of the beacon to the National Park Service.

With the area's reputation for fog, the lighthouse was an appropriate site for a horror movie. In *The Fog* (1980), a strange, glowing fog sweeps over a small coastal town in Northern California, bringing with it the ghosts of sailors who had died in a shipwreck there a century earlier. Ghosts would be bad enough, but these ghosts were out for revenge. The great cast (Adrienne Barbeau, Jamie Lee Curtis, Tom Atkins, Janet Leigh, and Hal Holbrook) and silly, so-so plot have earned the film a cult following.

Point Reyes is separated from the mainland by a neck of land bounded on the east by the San Andreas Fault, on the north by Tomales Bay, and on the south by Bolinas Lagoon. Although I feel as though I am still in North America, when I'm at the Point, geologically speaking I am on the Pacific Plate. East of the fault are the Franciscan rocks we saw on the Marin Headlands and Mount Tamalpais. Underpinning the Point Reyes peninsula is the same Salinian granite found near Salinas and Monterey 120 miles to the south. This means that the rocks have been carried more than 100 miles north by sideways movement along the San Andreas Fault. As if that weren't amazing enough, modern measurements of ground displacement suggest that Point Reyes has been moving northward at a rate of 0.5–1.6 inches per year relative to the North American Plate. If you do the math, at that rate it takes

4–13 million years to move the rocks 100 miles. Not only has the faulting moved the rocks northwest, it has uplifted, folded, and tilted them. The oldest rocks, the granites, are exposed as rocky outcrops at the lighthouse in the west and at Inverness Ridge in the east; between these outcrops is a gentle **syncline**, a bowl-shaped fold with the youngest rocks marked by grassy pastures lying in the center. The granite formed during the Cretaceous, between 100 and 80 million years ago, as a result of subduction and melting of the sea-floor and its overlying sediments.

Point Reyes Lighthouse, with its red roof, whitewashed lower walls, and massive upper glass windows, is perched precariously on a narrow spit of land jutting out into the storm-tossed Pacific. Waves crash against the rocks below in a never-ending din and boom, like a military fife-and-drum corps playing percussion with only a bass drum and the fife of seagulls calling as they hover in the wind.

About half a mile east of the lighthouse is Sea Lion Overlook. There is an interpretive panel and steps that go down to the beach. Here I see the salt-and-pepper granite that underlies the peninsula. As mentioned previously, this granite is also exposed at Tomales Point at the north end of the peninsula, and along Inverness Ridge, the north–south tree-covered ridge just west of the Bear Valley Visitor Center.

Above the granite, but not all visible in one place, is 11,000–14,300 feet of sedimentary rocks. The oldest **formation** is the Point Reyes Conglomerate, which occurs just above the granite at the Point Reyes Lighthouse. The **conglomerate** is made up of cobbles and boulders that eroded off the underlying granite and accumulated in deep marine channels and fans during the Paleocene, about 65 to 55 million years ago. The conglomerate appears to be identical to the Carmelo Formation found 100 miles south of here, on the east side of the San Andreas Fault near Monterey. This is another piece of evidence that the peninsula has moved north from its original location near Salinas. Point Reyes exists because the granite and conglomerate are hard, erosion-resistant rocks that can withstand the pounding of the ocean.

Above the conglomerates there is a time gap, an **unconformity** in the rocks that means new layers were not deposited for almost 30 million years or, more likely, were deposited and then completely eroded away. Above the missing rocks are Miocene (10–25-million-year-old) shallow-water marine deposits, the Laird Sandstone and Monterey Formation shale. They are found from Kehoe Beach to Tomales Point. When I hit the Monterey Formation with a hammer, there is a faint smell of decay, a "fetid odor." Petroleum geologists love finding rocks that have this smell, as it indicates that oil was generated in or moved through the rock, and a trace was left behind. We say that "it smells like money." The Monterey Formation happens to be the most important source of oil in California: organic matter in the black shale has generated an estimated 38 billion **barrels of oil**. Most of the Monterey Formation oil is found in the southern San Joaquin Valley and offshore from near Santa Barbara to the Los Angeles Basin.

The easily eroded Pliocene (2.5–5-million-year-old) sediments above the Monterey Formation form the lowland pastures between Inverness Ridge and Point Reyes. From oldest to youngest, these formations include the Santa Margarita Sandstone (exposed at Abbotts Lagoon), the Santa Cruz Mudstone (exposed at Duxbury Reef), and the Purisima Formation (at Drakes Bay). The Purisima Formation corresponds to identical rocks found in the Santa Cruz Mountains and Monterey Peninsula, once again indicating that they have moved north at least 100 miles. The Purisima Formation is the main rock that outcrops at our next stop, Drakes Beach.

From Point Reyes Lighthouse, return north on Sir Francis Drake Blvd to Drakes Beach Road on the right; turn right (southeast) and drive to the Drakes Beach parking area (38.028627, −122.961707), for a total of 6.8 miles (15 min).

STOP 7: DRAKES BEACH

This is a different kind of California gold. The broad sand beaches of Drakes Bay are not the typical dazzling-white sandy shorelines the state is known for. Drakes Beach exudes a warm golden glow. On this beach untrod by the multitudes that inhabit popular beaches from Malibu to San Diego, it is easy to encounter seals and sea lions lounging on the strand while seabirds circle and wheel overhead.

Drakes Beach is one of the most beautiful stretches of sandy beach in a state known for beautiful sandy beaches. The extensive arc of the south-facing shore is protected from the brunt of Pacific waves by the granite headland of Point Reyes. The wide beach is backed by abrupt and dramatic sandstone cliffs that rise 200 feet above the tide. The sea cliffs expose the Pliocene (about 2.6 million-year-old) Purisima Formation. These cream-colored siltstones and sandstones are reflected in the tide pools, and blaze golden in the setting sun.

But there's more than golden cliffs and sand here. The English pirate Sir Francis Drake made landfall in this bay to repair his ship, the *Golden Hind*, in June 1579. It is Drake whom the beach is named for. He was halfway through his voyage of exploration, which was really a series of raids,

Purisima Formation sea cliffs turned golden by the setting sun at Drakes Beach

authorized by Queen Elizabeth I, on Spanish and Portuguese shipping. At least partly as a result of Drake's raids, the Spanish began exploring the northern coast of New Spain to find a safer route from Mexico, the source of much gold and silver, to the spice islands of the East Indies.

Sixteen years after Drake visited, in 1595, the Spanish treasure ship *St. Augustine*, a 200-ton galleon loaded with porcelain and other goods from the Philippines, sank in Drakes Bay during a storm. Some of the surviving crew returned years later to salvage the ship. They found no evidence of the vessel. It wasn't until 1941, when archaeologist Robert Heizer was excavating local Native American sites, that traces of iron and Chinese porcelain were found. The ship's cargo of exotic treasure has never been found or salvaged. Partly as a result, the area became a National Historic Landmark in October 2012. Every now and then beachcombers still find bits of porcelain winnowed by the waves.

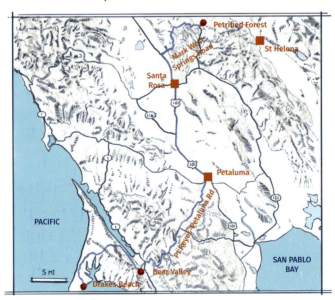

Map of the route from Point Reyes to Petrified Forest

MAYACAMAS MOUNTAINS

The coastal mountains of California highlight the variability of terrains and habitats found across the state. Crossing the first ridge, I notice a rather abrupt change from fog-shrouded coast to sun-drenched interior. Don't get me wrong—I love the fog on a hot day. But referring to the fog, there is a common quip often attributed to Mark Twain (although a Google search could not prove it): "the coldest winter I ever spent was a summer in San Francisco." So, leaving behind the fog-cooled rainforest at the coast, I enjoy the warmth afforded by the sunlit hills just a few miles inland. The Mayacamas Mountains, extending 50 miles northwest to southeast, are considered part of the Northern Inner Coast Ranges. The high point, 4,724 feet at Cobb Mountain, can be snow covered in winter, yet the average temperature of the range is surprisingly mild, in the 70s year-round. The range is covered by mixed evergreen forest and chaparral. The mountains were named after a Native American Yuki tribal village, Maiya'kma, just south of present-day Calistoga; the name appears on a Spanish land grant of 1841 for the Rancho Mallacomes y Plano de Agua Caliente. The *agua caliente* (hot water) probably refers to the hot springs and **fumaroles** that extend from Calistoga to The Geysers geothermal field.

Because of their location within the Coast Ranges near the boundary with the Central Valley, the Mayacamas Mountains consist of marine sedimentary rocks derived from both the Franciscan Complex and the Great Valley Sequence. The mountains themselves are a result of compression and uplift along strike-slip faults of the San Andreas Fault system, a relatively recent feature in this area. Thus these sediments are thought to have been uplifted and the mountains formed less than 15 million years ago.

The forces that brought these mountains into being are still active, and the comparatively young (geologically speaking) mountains will likely continue to rise as long as the

Pacific and North American Plates converge and grind past each other.

To get from Drakes Beach to Petrified Forest, return to Olema and drive northeast to 4100 Petrified Forest Road, Calistoga (38.555562, −122.638763), for a total of 67.2 miles (1 hr 40 min).

The drive to Petrified Forest takes me through Santa Rosa, a community of 178,000 (in 2020) nestled in the gentle and verdant Santa Rosa Valley. It is home to the Charles M. Schulz Museum (of Peanuts cartoon fame) and where the Santa Rosa plum was developed. The city is considered the gateway to the California wine country and Redwood Coast.

It's 2019 and I'm driving through Coffey Park, one of the northern neighborhoods of Santa Rosa, when suddenly the landscape changes to a scene of utter destruction, like the aftermath of Sherman's March to the Sea. Everything had burned: concrete home pads and burned-out cars litter the side streets. Nothing was left standing except chimneys and a few charred trees. In October 2017, an electrical malfunction next to a residential structure during the seasonal Diablo winds caused an inferno that engulfed 36,000 acres and killed at least twenty-two people. Known as the Tubbs Fire, it destroyed more than 5,643 structures, half of them homes in Santa Rosa. While almost all of the devastation has been rebuilt, this serves as a reminder of the vulnerability of this region to the violence of wildfire.

The hot, dry winds out of the north and east always come in October, as the seasons shift in California. When the 40–60 mph winds arrive at the end of the dry season, all it takes is a spark to set off a firestorm. Our local fire chief went to fight this particular blaze and described flames that the wind blew flat across six lanes of freeway. The fire burned all the way to our next stop, Petrified Forest, which had to close for a few years to rebuild. In an ironic twist, the only reason the **petrified** trees are there is because they were buried in ash.

STOP 8: PETRIFIED FOREST, CALISTOGA

Wait. A petrified forest in California? Isn't that in Arizona?

Well, yes, there *is* a petrified forest in Arizona, but don't confuse this petrified forest with its more famous Arizona cousin. Arizona's Petrified Forest National Park contains trees that are Triassic, about 218–225 million years old. All the trees found in the national park are extinct members of the *Araucaria* genus, evergreen conifers that today include the Norfolk pine. California's Petrified Forest, by contrast, contains much younger, 3.2–3.4-million-year-old (Pliocene) redwood trees of the genus *Sequoiadendron*. Yup, they are sequoias.

The trees were preserved and petrified when they were buried in ash from a late eruption of the Sonoma Volcanics or an early eruption in the Clear Lake Volcanic Field near Mount Saint Helena, 8 miles north of here. The volcanic ash became the unlikely savior of these redwood trees. After they were buried beneath many layers of ash, groundwater slowly worked its magic. Silica in the volcanic ash slowly dissolved into the groundwater and was redeposited in the logs, where it replaced the original wood one cell at a time.

One of many petrified redwood trees at Petrified Forest

The stone trees in this petrified forest look amazingly wood-like, as if they had just fallen on the ground. These trees are the direct ancestors of the giant sequoias found today only in the Sierra Nevada from Sequoia National Park to Yosemite. It's like meeting the grandparents of some of the largest living beings on Earth. And being sequoias, they are of course the world's biggest petrified trees.

California's Petrified Forest was publicized in 1870 by a homesteader, Charles Evans, known to his friends as Petrified Charlie. Charlie was able to get the most famous American **paleontologist** of the time, Othniel Marsh, to come look at the stone trees in 1871. Marsh is the one who determined that the trees are an extinct variety of redwood. Sometime later the novelist Robert Louis Stevenson (*Treasure Island, The Strange Case of Dr. Jekyll and Mr. Hyde*) honeymooned on the slopes of nearby Mount Saint Helena and mentioned the Petrified Forest in his 1883 book *The Silverado Squatters*.

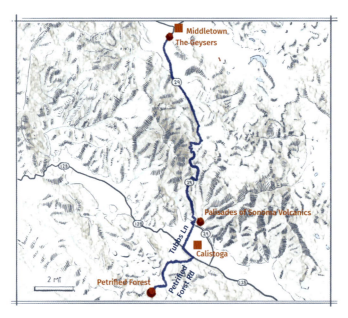

Map of the route from Petrified Forest to Palisades of the Sonoma Volcanics and The Geysers

Since 1914 the property has been owned by the family and descendants of Ollie Bockee. The family has worked with a number of naturalists and universities, in particular UC Berkeley, to further the knowledge of this ancient, well-preserved redwood forest.

As the forest is on private property, there is an entrance fee. The forest is located at 4100 Petrified Forest Road, Calistoga. See the website for hours and admission: www.petrifiedforest.org.

Our next stop is The Geysers, one step closer to the source of the volcanic ash that buried these trees. Out of the frying pan, into the fire, so to speak.

A few miles off the main route of this tour are the world-famous fumaroles, hot springs, and geothermal energy plants at The Geysers. The folks who operate the field are happy to provide a free tour, but they would like guests to register in advance. Sign up online for a geothermal power plant tour led by the Calpine staff. Tours start at the Calpine Geothermal Visitors Center at 15500 Central Park Road in Middletown. View the schedule and register for a tour at geysers@calpine.com or http://geysers.com/Visitor-Center-and-Tours. You can also just arrive and wander through the visitor center.

From Petrified Forest, continue northwest to the Calpine Geothermal Visitor Center, 15500 Central Park Road, Middletown (38.745252, −122.621570), for a total of 20.2 miles (33 min).

If you prefer to skip The Geysers, as you leave Petrified Forest, continue northwest to CA-29 and drive to a pullout on the right (south) side of the road, Palisades of the Sonoma Volcanics (38.608253, −122.594539), for a total of 6.3 miles (12 min).

STOP 9: THE GEYSERS

I was expecting to see the California version of Old Faithful, the Yellowstone **geyser** that spurts steaming-hot water

180 feet in the air once every hour. It so happens that there are no geysers at The Geysers. The name was suggested to early visitors by the fumarolic activity and steam rising from hot springs. But that doesn't make the area any less interesting. The first known encounter by a settler was in 1847, when William Bell Elliot, while tracking a grizzly bear, found steam vents, or fumaroles, along Big Sulfur Creek. He happened upon a startling sight: puffs of sulfur-laden steam emerging from the hillsides. The sulfur lends the valley an aroma of rotten eggs. Standing amid this smelly, surreal scene, Elliot couldn't help but think he had stumbled on the gates of hell itself. It is still known as Devil's Canyon.

Since then, the area has transitioned from a natural wonder to a resort, and later to a source of clean geothermal energy.

Devil's Canyon, The Geysers, around 1868 (US Library of Congress's Prints and Photographs division, digital ID ppmsca.09994. https://commons.wikimedia. org/wiki/File:Carleton_Watkins,_California_ geysers,_Devil%27s_Canyon,_ca._1868.jpg)

Archibald Godwin developed a spa here, the Geysers Resort Hotel, in the years 1848–1854. People from far and wide flocked to the hot springs to "take the waters." Hot-spring waters were believed to cure such diseases as arthritis and to lower blood pressure, strengthen bones, and relax muscles. Hot-spring spas were all the rage in the 1800s, and still are in many parts of the world. The upscale resort at The Geysers attracted famous tourists, including Ulysses S. Grant, Mark Twain, and Theodore Roosevelt. Over time the resort declined in popularity, and it was partly destroyed by landslides and fires. The last remnants of the once-glamorous Geysers Resort Hotel were eventually demolished in 1980, marking the end of an era.

The area next became a source of environmentally friendly energy. Geologists sometimes get a bad rap for being involved in the production of carbon-based energy. One of the remarkable aspects of geothermal energy is that it is kind to the environment. Geothermal energy is considered "green" because it produces no greenhouse gases, doesn't produce radioactive waste, and doesn't flood valleys.

The Geysers steam field is located on the southwest edge of the Clear Lake Volcanic Field. Evidence of volcanic activity is still very much alive in the form of steaming hot springs. Groundwater at The Geysers is heated by a near-surface **magma chamber** known as "the felsite." The felsite is between half a mile and four miles below the surface, and began cooling there about a million years ago. Related rocks at the surface include 1.6-million-year-old black basalt lava flows at Caldwell Pines, 1.1-million-year-old **rhyolite** lava flows at Cobb Mountain, and 3.4-million-year-old volcanic ash at Petrified Forest. Rhyolite is a light-colored and fine-grained lava with the composition of granite. These volcanic units were deposited over the Franciscan Assemblage. The Franciscan in turn sits over the felsite. In fact, a fractured sandstone in the Franciscan Assemblage holds most of the steam that is produced to make electricity.

The rocks that hold the steam are fractured and altered graywacke sandstone. The steam comes from superhot groundwater, with temperatures 500°–750°F. But because the water is under immense pressure, it doesn't flash to steam until the pressure is released. Releasing the pressure is like popping the cork on a bottle of champagne: once the pressure is released, the liquid comes frothing out of the container. The pressure is released as the water moves upward to the surface, either naturally or in geothermal wells.

Geothermal wells are drilled down into the superhot water zone to produce steam, which in turn drives turbines to generate electricity. The first geothermal wells at The Geysers were drilled by the Geysers Resort Hotel. Starting in 1921, the steam-powered generators produced 250 kilowatts per hour. Because 1 kilowatt is only enough energy to power a 100-watt light bulb for 10 hours, this energy was used mainly to light the resort.

Sensing a business opportunity, power companies moved in and began large-scale drilling in 1955. In 1960, Pacific Gas and Electric began producing electricity from an 11-megawatt power plant. This was roughly enough electricity to power the demands of 8,250 homes at once.

There was one potential glitch in tapping this source of green energy, though. By 1999 the produced steam was depleting the groundwater and reducing the pressure in the underground steam reservoir. So companies began recycling the produced steam by condensing it at the power plants and reinjecting it back into the hot rock layers. They also began to inject up to 18 million gallons per day of treated wastewater from local communities into the field to maintain the groundwater system.

So far so good, right? As with many industrial processes, there were some unintended consequences. Reinjecting water caused shallow, man-made earthquakes. That is because the water acts as a lubricant, allowing faults that are under stress to slip and rocks to break. Most of the quakes were too small to be felt, around magnitude 1 and 2 on the Moment magnitude earthquake scale. But a few magnitude 3 and rare magnitude 4 quakes were strong enough to rattle nearby communities. Public outcry led to more careful management of water injection, which has decreased the number of earthquakes to acceptable levels over the past 25 years.

The Geysers' role in sustainable energy production serves as a reminder that even the greenest of technologies require careful consideration and management.

THE SONOMA VOLCANICS

In contrast to the relatively recent Clear Lake Volcanic Field and its expression at The Geysers, the Pliocene-age Sonoma Volcanics erupted from multiple vents over a period of several million years between 2.5 and 8 or 9 million years ago. These volcanics are mainly light-colored volcanic ash, medium-dark **andesite**, and black basaltic lava flows. The lavas and ash cover large parts of the Howell and Mayacamas Mountains, and occupy most of the ridges on both sides of Napa Valley. They are an important component of the highly fertile soil in the Napa and Sonoma winegrowing region.

Perhaps not surprisingly, the town of Calistoga at the north end of the Napa Valley is famous for its hot springs, sparkling mineral pools, mud baths, and, until recently, sparkling spring water. An early settler in the area, Giuseppe Musante, came across an effervescent, mineral-rich water after drilling a well. He sold this water at his downtown Calistoga candy shop starting in 1920. He would later go on to found the Calistoga Mineral Water Company. In 2022, the company, renamed Crystal Geyser, sold 85 million bottles of the sparkling water. The plant closed in 2024 for economic reasons.

To get from The Geysers to the Palisades of the Sonoma Volcanics, drive 13.8 miles (22 min) south on CA-29 to the Palisades of the Sonoma Volcanics (38.608253, −122.594539) and pull over on the left (south) side of the highway.

STOP 10: PALISADES OF THE SONOMA VOLCANICS

The Palisades are magnificent ridgetop cliffs that dominate the landscape around Calistoga. They are a natural wonder, towering formations that are a striking representation of the Sonoma Volcanics. Rising to heights of 600–700 feet, the Palisades are composed of alternating layers of volcanic mudflows, welded ash, an array of volcanic rock fragments, and columnar andesite. This distinctive rock is the result of volcanic eruptions that took place near here around 3.4 million years ago.

A remarkable aspect of the Palisades is the display of **columnar jointing**, a peculiar feature of lavas that occurs when molten rock cools slowly and contracts. As it solidifies, it fractures in a distinct hexagonal pattern, forming a series of polygonal columns that resemble towering organ pipes. These are the same features that have captured the imagination of visitors from all over the world at Devils Postpile and Devils Tower National Monuments. Whenever I see these, I think of Richard Dreyfuss's character at the dinner table, sculpting Devils Tower out of mashed potatoes in *Close Encounters of the Third Kind*.

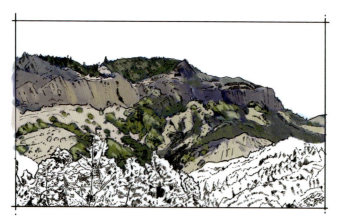

View east to the Palisades of the Sonoma Volcanics from the CA-29 overlook

The columns, standing tall above the valley, along with the hot springs in the valley, are a reminder of the fiery forces that shaped and continue to shape our world.

To get to Glass Mountain, continue driving southeast on CA-29 to Silverado Trail. Take Silverado Trail to the Glass Mountain roadcut (38.530213, −122.490713), for a total of 9.7 miles (15 min).

STOP 11: GLASS MOUNTAIN

I love obsidian. This black volcanic glass can't be found just anywhere. It takes an edge better than any other rock, and has been prized by Native Americans for knives, spear points, scrapers, and arrowheads. Before European contact, it was traded long distances among the various tribes. To me, it's a rare, beautiful rock.

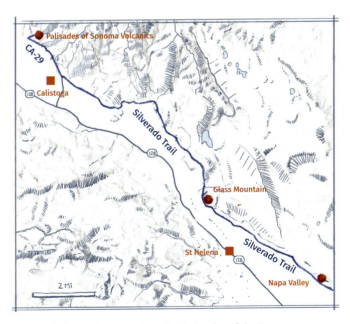

Map of the route from the Palisades of the Sonoma Volcanics to Glass Mountain and Napa Valley

Glass Mountain is made up of volcanic ash of the Sonoma Volcanics. In this roadcut are bits of pumice, a frothy lava that has so many gas bubbles that it is the only rock that floats. The roadcut also has chunks of obsidian, a glassy rock with a composition similar to granite. Obsidian forms when a lava flow cools exceedingly fast, preventing the growth of large crystals. This usually happens when a river, lake, or pond quenches silica-rich rhyolite lava. The obsidian in this ash layer is black due to impurities such as iron. At Glass Mountain, pure black obsidian weathers out of the lighter-colored ash in this roadcut like raisins falling out of a crumbling oatmeal cookie. I have found that I rarely need a rock hammer: there are loose chunks of the black glass lying all around.

I notice that rock bits in the outcrop are poorly sorted; that is, there are fragments of many sizes and shapes all found together. To me this suggests a **pyroclastic flow**, a superheated gas cloud that rolls down a mountainside as a result of explosive volcanism, which mixes all manner of ash, lava, and older rocks. This is the type of explosive eruption that caused a

Glass Mountain roadcut, Silverado Trail. Obsidian makes up the darker chunks in this massive ash flow.

The columns, standing tall above the valley, along with the hot springs in the valley, are a reminder of the fiery forces that shaped and continue to shape our world.

To get to Glass Mountain, continue driving southeast on CA-29 to Silverado Trail. Take Silverado Trail to the Glass Mountain roadcut (38.530213, −122.490713), for a total of 9.7 miles (15 min).

STOP 11: GLASS MOUNTAIN

I love obsidian. This black volcanic glass can't be found just anywhere. It takes an edge better than any other rock, and has been prized by Native Americans for knives, spear points, scrapers, and arrowheads. Before European contact, it was traded long distances among the various tribes. To me, it's a rare, beautiful rock.

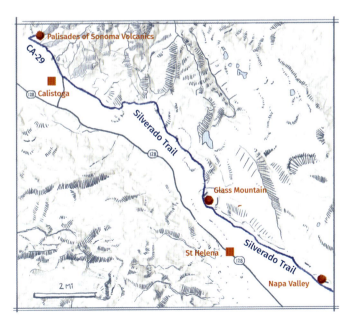

Map of the route from the Palisades of the Sonoma Volcanics to Glass Mountain and Napa Valley

Glass Mountain is made up of volcanic ash of the Sonoma Volcanics. In this roadcut are bits of pumice, a frothy lava that has so many gas bubbles that it is the only rock that floats. The roadcut also has chunks of obsidian, a glassy rock with a composition similar to granite. Obsidian forms when a lava flow cools exceedingly fast, preventing the growth of large crystals. This usually happens when a river, lake, or pond quenches silica-rich rhyolite lava. The obsidian in this ash layer is black due to impurities such as iron. At Glass Mountain, pure black obsidian weathers out of the lighter-colored ash in this roadcut like raisins falling out of a crumbling oatmeal cookie. I have found that I rarely need a rock hammer: there are loose chunks of the black glass lying all around.

I notice that rock bits in the outcrop are poorly sorted; that is, there are fragments of many sizes and shapes all found together. To me this suggests a **pyroclastic flow**, a superheated gas cloud that rolls down a mountainside as a result of explosive volcanism, which mixes all manner of ash, lava, and older rocks. This is the type of explosive eruption that caused a

Glass Mountain roadcut, Silverado Trail. Obsidian makes up the darker chunks in this massive ash flow.

massive tsunami and wiped out the islands around Krakatoa, west of Java, Indonesia, in 1883. I would *not* want to have been standing here when this flaming ash flow erupted 2.5 million years ago.

Having said that, this volcanic ash played a significant role in enriching the soils of the Napa Valley, contributing to its fertility and making it a prime region for viticulture. Which is one of the reasons I am here now.

To get to the Napa Valley stop, drive 4.9 miles (7 min) southeast on Silverado Trail to a pullout on the right (38.492562, −122.417732).

NAPA VALLEY AND THE *TERROIR* OF WINE

My encounters with the Napa Valley began some forty years ago. My wife and I would leave our young kids with her parents and escape to Yountville. We spent our mornings tasting wine, would have a picnic lunch at a local park, then spent the afternoons lying in the warm sun by the hotel pool. There is something enchanting about this valley, the green of the vines so bright they absorb the light and throw it back at you. The allure is only enhanced by thoughts of enjoying a refreshing glass of wine. The Napa Valley is the perfect mix of sun and warmth and water and soil. It is wide and flat and rimmed by gently rolling hills that rise hundreds of feet above the valley floor. South- and west-facing slopes are sunbaked and mostly covered with chaparral and grass; north- and east-facing slopes are covered in oak and pine woodland. The valley itself is a scene of intense agriculture, a patchwork quilt of vineyards that fill the valley and creep hesitantly up the surrounding hillsides. This is expensive property, the playground of old money and new wealth.

For all these reasons, the Napa Valley has become synonymous with fine wine. It is regarded as one of the best winemaking regions in the world, and was named the state's first American Viticultural Area (AVA) in 1981. But winemaking has been fundamental to this region as far back as the late 1700s, when Spanish Franciscan missionaries planted the first vines of *Vitis vinifera* for use in the sacrament. It wasn't until 1839 that a gentleman by the name of George Yount introduced modern grape cultivation practices. Yount planted vines on a grant of land from the Mexican government. Soon he was overseeing a number of vineyards in Northern California. Others followed his lead. Yount's first neighbor, Edward Bale, planted the old missionary grapes; another neighbor, Samuel Brannan, a short time later purchased a large tract and planted more than 100 acres of vines. Brannan named the tract Calistoga.

John Patchett is generally recognized as starting Napa's first "official" vineyard and winery. He began planting vines in 1854, four years after California gained statehood. By 1857, Patchett was producing wines with the help of his winemaker, Charles Krug. Winemaking was so undeveloped in California that there were no wine presses. Patchett and Krug had to use a cider press. The first review of Patchett's wine, published in *California Farmer Magazine* in 1860, said: "The white wine was light, clear and brilliant and very superior indeed; his red wine was excellent."[7]

Krug would go on to start his own winery in 1861, and it is the oldest continuously operated winery in the region. Around the same time, Joseph Osborne planted vines on an 1,800-acre tract he called Oak Knoll. Hermosa Vineyards was started by Hamilton Crabb in 1872, and by 1878 he was one of the largest landholders in the Napa Valley. Some of the earliest wineries are still around, including Beaulieu, Beringer, Chateau Montelena, Far Niente, Inglenook, and Schramsberg. These early wineries imported a number of grape varieties, including Cabernet Sauvignon, Zinfandel, Charbono, Malbec, and others.

There was apparently enough of a demand that these early successes sparked a wave of investment and development. By

1890, almost two hundred wineries had been established, and over 18,000 acres of vines had been planted. Then, like all booms, the wine boom came to an end when the industry suffered a series of setbacks. Starting in the 1860s, an infestation of phylloxera, an aphid-like insect that feeds on the roots of vines, devastated the valley's vineyards. By 1900, the bugs had destroyed over 90 percent of Napa Valley vines. Then, at the turn of the century, a surplus of grapes forced down prices, putting more vineyards out of business. The coup de grâce came in 1920, when Prohibition killed many of the remaining commercial vineyards and wineries. A few properties continued operations by going back to their origins and making sacramental wine for churches.

By 1933, Prohibition had been repealed, and Napa Valley vineyards slowly began to recover. Wineries reopened, and an ever-increasing demand for wine buoyed the industry. The Napa Valley Vintners Association was formed in 1944; today it represents over 550 wineries. These days it seems as though every celebrity who can afford to buy some land wants to grow grapes or make their own wine. They range from Mario Andretti and Francis Ford Coppola to John Legend, Joe Montana, even Nancy Pelosi and Yao Ming.

It wasn't until 1976 that Napa wines became world famous. In that year, the historic Judgment of Paris saw Cabernet Sauvignon and Chardonnay from California pitted against Bordeaux and Burgundy in a blind tasting. The judges gave top honors to two Napa Valley wines, Chateau Montelena's Chardonnay and Stag Leap's Cabernet Sauvignon. The winegrowers of France and Italy began to take California wines seriously. As they say, the rest is history.

A unique combination of soil, climate, and topography has blessed the vines of the Napa Valley. The French term *terroir* (pronounced "tare-wahr") describes the role of geology, soil, and climate in producing the best wine (or, for that matter, coffee or tea). The term is meant to encompass the totality of the environment, or sense of place of the wine.

Bedrock being the parent material for soil, it provides much of the original nutrients. The minerals themselves are flavorless and odorless. Yet minerals such as feldspars, **micas**, amphiboles, and pyroxenes break down through chemical weathering and release the nutrients that plants need to grow. Another source of nutrients is organic matter. Mature soils on gentle slopes accumulate organic matter from the breakdown of dead vegetation, whereas steep rocky slopes with thin soils do not. Plants growing on the soil, whether grasses, shrubs, or trees, eventually die, decompose with the help of bacteria and fungi, and get mixed back into the soil by earthworms, moles, and other burrowing creatures. Landslides, rivers, and floods do the heavy mixing.

Soil texture is as important as nutrients when it comes to watering the vines. Texture includes the soil's **porosity** (spaces between grains) and **permeability** (how easily water moves through it). The absence or presence of inert minerals such as quartz contributes to the soil's ability to hold moisture or drain well. The roots of vines generally appreciate well-drained soils.

The bedrock of the Napa Valley is mainly the Franciscan Complex, which I described earlier as a mixture of basalt, chert, and sandstone. The Great Valley Sequence of sandstones and shales is found on the east side of the valley and contributes to the mix. Weathering volcanic rocks on the ridges east and west of the valley complete a blend that both provides nutrients and drains well.

Farmers around the globe know that volcanic ash develops into great soils. The small grains with large surface areas weather rapidly and provide lots of mineral nutrients to the soil, including iron, magnesium, potassium, and aluminum. These minerals are natural fertilizers. Volcanic soils are also light and fluffy compared to other soils, and because of this they easily retain moisture. The area around the Napa Valley contains rich soils because eruptions in the Clear Lake Volcanic Field 30 miles north of here sent ash into the air that settled out in the valley. The soils also benefit from weathering of the Sonoma Volcanics that outcrop around the valley rims. The Sonoma Volcanics are 2.5 to 8 or 9 million years old, and eruptions from the Clear Lake Volcanic Field occurred mostly between 10,000 and 2 million years ago. Both have had ample time to weather into soils in this alternating dry and humid, temperate climate.

Climate determines the availability and timing of sun and moisture. Temperate climates, because of their seasonal variations in moisture and plant growth, have some of the best-developed and most fertile soils. The temperate Mediterranean climate in the Napa Valley provides plenty of sunshine to encourage grape growth.

Likewise, topography is crucial to soil development. Steep slopes don't develop thick soils; rather, the soil moves downslope and accumulates in low, flat areas. Flat and gently rolling topography is where the thickest soils develop. The Napa Valley has both flat and rolling topography, so generally thick soils.

Slope aspect—the direction the slope faces—plays a role in retaining soil moisture, and soil moisture promotes the weathering of rocks. Plants need moisture, but too much dampness and root rot sets in. Some grapes, usually those used to make red wines, prefer the drier and warmer south-facing slopes, whereas others (white wine grapes) thrive on the cooler and moister north-facing slopes.

It is well known among growers that soil composition and moisture affect the character of the grapes. Grapes grown on valley floors, in deep, fertile soils, produce large and abundant grapes, so growth must be managed to concentrate the flavor. Vines grown on hillsides in rocky soil have to struggle to survive; they create small crops with grapes that have highly concentrated flavors.

The Napa Valley itself is bounded by faults. It has been dropped down relative to the hills on either side, like a long trench that has filled with alluvial soil. The soil, combined

Typical Napa Valley vineyards

with a mild climate, abundant sun, and sufficient rain, produces world-class wines. It is reported that red wine grapes grown on weathered basalts have sharp flavors that require aging to mellow. White wine grapes grown on serpentinite allegedly have a more acidic taste. Since I am here, I plan to confirm these observations and perhaps add a few of my own.

The vineyard I describe here happens to be along our way. My intent is not to promote any particular winery but to provide an example of a vineyard experience. There are many vineyard tours, some more expensive and some less so. A quick Google search turns up dozens of pages of wineries.

If you stop at Quintessa, make reservations well in advance for busy summer and weekend tastings. For questions or reservations, visit its website at quintessa.com.

To try one of many vineyard experiences, continue south from Glass Mountain for 0.6 miles (about 4 min) on Silverado Trail to Quintessa Vineyards (38.487141, −122.410911).

STOP 12: NAPA VALLEY AND QUINTESSA VINEYARDS
Winemakers have come to Napa from around the world, but mostly from the winegrowing regions of Spain, Italy, and France. An exception would be the founders of Quintessa Winery, the Chileans Agustin and Valeria Huneeus. Having recently visited Chile with the intention of learning about the country's wines, I can vouch that they have world-class wineries. The first vines at Quintessa were planted in 1990. Due to the size of the property and geographic features in the valley, the winery contains a number of exposures, slopes, and soil types. This was one of the last large "virgin" Napa Valley floor properties available at the time of purchase: it had never been planted in grapes. The vineyard contains a mix of hillside and valley floor vines. Quintessa's first vintage was released in 1994.

Quintessa Vineyard has a total of 160 acres planted with Cabernet Sauvignon, Merlot, Cabernet Franc, Petit Verdot, and Carménère grapes. According to the vineyard's website, the grapes are harvested early in the day; sorted and fed into oak, stainless steel, or concrete tanks; and fermented. A 17,000-square-foot wine cave was tunneled into the hillside beneath the hospitality center. At maximum capacity, about 3,000 barrels from two vintages can be stored there. After fermentation, the wine is aged for up to two years in French oak. After final blending, the wine is bottled and aged another year before it is released to the public. Uniquely in Napa Valley, Quintessa specializes in only one wine each year.

Born in Chile, the Huneeus family are wine industry veterans. Valeria is a microbiologist and viticulturist. Agustin inherited a fishing business from his father, but took advantage of an opportunity to invest in Viña Concha y Toro, a small winery on the outskirts of Santiago. He helped build Concha y Toro into Chile's largest wine producer and exporter. In 1960, at age twenty-seven, Agustin became the winery's CEO. In the early 1970s, he worked at Seagram's, and within a few years he managed its worldwide operations. In 1999, Agustin founded Huneeus Vintners, which owns Quintessa and other brands, and has properties in California and Chile.

If you sign up for the estate tasting, after checking in, the tour begins with a walk up the hill behind the winery to an overlook offering expansive views of the valley floor, Quintessa's vineyards, and the large pond below. The wine tasting takes place at a private table located in the interior of the stone building beneath glass skylights. This is an intimate room with a few small tables. A wine guide supervises the experience and explains in great detail each of the wines and how they are made. Vineyard samples are provided so guests can taste the individual components making up the final wine. Then a current vintage is offered along with a meal specially created by the winery chef. Each part of the meal is paired with a different wine.

Fine wine and good food are hallmarks of the Napa experience, drawing tourists from around the world. Napa and the surrounding valleys have a well-earned reputation for the quality of their climate, soil, and wine grapes. The Napa Valley may get all the headlines, but the **Great Valley**, or Central Valley, lying just to the east, is the true agricultural colossus.

CHAPTER 2

The Great Valley

From the Napa Valley there is one more range to cross, the eastern Coast Ranges, to get to the Central Valley, popularly known as the Great Valley. Cresting the last rise, I stare out over the valley. The still air is suddenly warm, not at all like the cool coastal breezes I am used to. The valley is laid out before me, a seemingly endless expanse of orchards and farmland backed by barely seen, hazy distant mountains. Thunderheads caused by hot air rising on the far-off slopes mark the Sierra Nevada. It is hard to tell where the mountains end and the cumulus clouds begin.

This geological basin is as flat as a pancake except for the rare volcanic butte. I like to visualize in my mind's eye what it could have looked like 5 or 10 million years ago. It's not too hard to imagine that this flat expanse was, until recently, submerged under the sea. As the basin filled with sediment, the sea became progressively shallower, until it was full to the brim. In places there are as much as 43,000 feet of sedimentary rock beneath the surface, mainly marine muds interlayered with sandstones eroded from volcanic and granitic rocks to the east. The eastern shore of the basin was effectively the west coast of North America, and the Coast Range would have been a string of islands, an offshore archipelago. As the basin filled from east to west, the rivers flowing from the distant mountains began to merge into the two large streams that now drain the valley, the San Joaquin

River flowing north, and the Sacramento River flowing south. They join in the fittingly named Sacramento–San Joaquin River Delta just southwest of Sacramento and flow into the large estuary that is San Francisco Bay.

THE GREAT VALLEY SEQUENCE

Had I been standing in the foothills of the Sierra 70 million years ago, a few million years before the meteor impact that ended the age of dinosaurs, looking west I would have seen nothing but ocean all the way to the horizon. The volcanic island arc that had been offshore 145 million years ago, in the Late Jurassic, had long since been fused onto the continent. Behind me, rising perhaps 10,000–15,000 feet, would have been a range of active volcanoes, the current volcanic arc. Even when they weren't popping off, there would have been steam vents and hot springs and fumaroles. Lava flows and volcanic ash deposits would have covered the landscape.

The rich soils of the Central Valley are the contribution of the Sierra Nevada, that magnificent mountain range with both an ancient and modern history. In the distant past they were a volcanic range; by 20 million years ago they had eroded almost to sea level. More recently, over the past 5–10 million years, they became a fault-bounded uplifted block of granite. These mountains contributed the sand and mud that pushed back the ocean and generated the bountiful soils of the region. And the water. Between rain and the winter snowpack, these mountains provide the water that irrigates 6 million cultivated acres. Not only is California the top food-producing state in the US, but according to the California Department of Food and Agriculture, "California agriculture is a $49 billion industry that generates at least $100 billion in related economic activity." Apparently, all it takes is sun, water, good soil, and hard work. The rocks filling the Central Valley are mainly deep marine muds that accumulated off the Jurassic and Cretaceous (66–200-million-year-old) west coast of North

America. Turbidity flows brought sand and conglomerate from the eastern mountains. **Brachiopods** (shellfish that look like clams, but aren't), **pelecypods** (clams), **gastropods** (snails), and ammonites (related to the coiled nautilus) have been found in these marine rocks. Some vertebrates, including fish and marine reptiles, have been found in the uppermost, youngest parts of the sequence.

At the time these rocks were being deposited, subduction was the dominant process along the west coast of North America. The Sierra Nevada was an active volcanic arc, much like the Cascade Range is today. Sediments that eroded off the Sierra were deposited in the ocean that is now the Central Valley. Farther west was an oceanic trench where the east-moving Farallon Plate (oceanic crust) was diving beneath the west-moving North American Plate. As I described earlier, as the oceanic plate moved down into the mantle, the

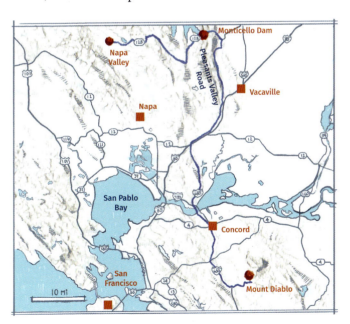

Map of the route from Napa Valley to Monticello Dam and Mount Diablo

sediments carried by the plate, as well as part of the upper surface of the plate itself, were scraped off and pushed westward along a series of thrust faults. These rocks became the Franciscan Terrane.

Before 30 million years ago, the North American and Farallon Plates were converging, moving toward each other. After about 25 million years ago, they were moving side by side, slipping past one another along the San Andreas Fault system in what's known as a transform margin. Because ocean crust was no longer going under the continent and melting, the Sierran volcanic arc went extinct. In its place the San Andreas system of faults became active. As the two plates slid past each other, there was a certain amount of compression between them. This caused tilting of the Great Valley Sequence to near-vertical along the west side of the Central Valley. A large section of this is in full display at the next stop.

To get to Monticello Dam, drive southeast on Silverado Trail to CA-128/Sage Canyon Road; turn left (east) on CA-128 and drive to the Monticello Dam parking area (38.512253, −122.103428), for a total of 27.2 miles (40 min).

STOP 13: THE GREAT VALLEY SEQUENCE AT MONTICELLO DAM

A magnificent exposure of the Great Valley Sequence occurs along Putah Creek and at Monticello Dam. The 304-foot-high Monticello Dam was built to hold back the waters of Lake Berryessa so as to prevent flooding, provide water for irrigation, and create a place to jet ski in the summer. The reservoir provides irrigation water to Solano and Yolo Counties. Lake Berryessa was named after José Jesús Berrelleza, a Basque rancher who was granted Rancho Las Putas in this area in 1843. Rancho Las Putas was named after the putahs, or suckerfish, found in Putah Creek. Construction of Monticello Dam began in 1953, and the lake was filled

Outcrops of the Venado sandstone interval, Great Valley Sequence, at Monticello Dam. These layers were originally horizontal, with the top being to the left. Notice how the layers range from thick to thin and repeat this pattern. The coarse grains came out of the turbid suspension first, then finer and finer grains, until a thin layer of mud settled out on top.

by 1963. Prior to filling, the valley had some of the finest agricultural land in the state. The town of Monticello was drowned beneath the reservoir, and the story of the loss was documented and made famous by a photo essay, *Death of a Valley*, by Dorothea Lange and Pirkle Jones.

Lake Berryessa fills a valley bounded on both sides by rocks of the Great Valley Sequence. It is an important source of groundwater recharge for the surrounding region. Water from the lake percolates downward through the porous layers and replenishes groundwater that has been removed from these layers beneath the Central Valley by pumping in water wells.

As you drive along the lake, there are several excellent roadcut outcrops of turbidite deposits consisting of dark shales and interbedded thin sandstones. These are submarine landslide deposits within the Great Valley Sequence.

Pull over at the parking lot at Monticello Dam. The high ridges here are 93-million-year-old thin-to-thick sandstones and interbedded shale and conglomerate layers of the Venado interval of the Cortina Formation. The Venado sandstone interval is another one of those **turbidite fan deposits**—that is, sands and muds that ran down a submarine valley and fanned outward onto the deep ocean floor. Like a delta at the mouth of a river, but under water. Gastropods, a type of mollusk in the snail clan, as well as the burrows of shrimp-like sea creatures, have been found in the Venado interval.

State Route 128 winds its way through Thompson Canyon, one of the best locations to view the Great Valley Sequence. The eastern extensions of these rocks lie buried beneath the cornfields and fruit orchards of the Central Valley. As I drive east along Putah Creek, with steep walls of rock rising on both sides, the layers of the Great Valley Sequence go from vertical at Monticello Dam to near horizontal at the mouth of the canyon. This is because the rocks in this valley are undeformed to the east, but were pushed upward at the west end by faults bounding the Coast Ranges.

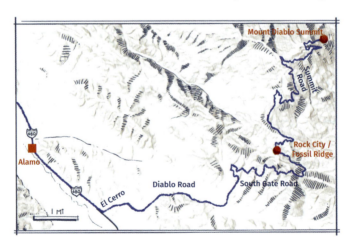

Map of the route to Rock City/Fossil Ridge and the summit of Mount Diablo

To get to Mount Diablo from Monticello Dam, drive to I-80 and head west to US-680; drive south on US-680 to Exit 40; turn east onto El Cerro Blvd; continue straight onto Diablo Road and drive to Mt. Diablo Scenic Blvd; turn left (north) on Mt. Diablo Scenic Blvd to South Gate Road and drive to Rock City parking area on the left (37.848402, −121.931814), for a total of 67.8 miles (1 hr 30 min).

THE DEVIL'S THICKET

Mount Diablo is an isolated peak that sits at the boundary between the Coast Ranges and Central Valley. Because it is surrounded by rocks of the Great Valley Sequence and provides a glimpse of that sequence that is buried east of here, I'm calling it part of the Great Valley. The peak rises to 3,849 feet and contains rocks from almost every episode that formed Northern California. It is located within Mount Diablo State Park. Mount Diablo was designated a state park in 1921, and over the years the park has grown to over 20,000 acres. Today the mountain is a favorite of cyclists; please watch for them, as there are only a limited number of turnouts. It's not so bad when the cyclists in their skin-tight nylon are going up, pumping the pedals of their custom bikes. But watch out on the way down: if you aren't going fast enough, they *will* pass you doing 50 miles an hour. The mountain is so well known among bikers that there is a Mount Diablo Challenge bike race each year on the first Sunday in October. Proceeds go to local land conservancy programs.

It is interesting to see how some Native American origin stories reflect the geology of the region. According to the Tachi Yokuts of the southern San Joaquin Valley, the Ancient Ones, portrayed as animals, formed the land from the bottom of the ocean. Turtle brought mud from the seafloor to create the land. According to the geologic origin story, the Central Valley was indeed seafloor mud and sand that has since risen above sea level.

In the origin stories of some local Indigenous groups, Mount Diablo is the center of creation. According to Miwok and Ohlone creation stories, Coyote, the creator, and his assistant Eagle-man, fashioned people and the rest of the world from the two islands of Mount Diablo and Mount Tamalpais.

The story of how the mountain got its Euro-American name is closely bound to the history of the region. The Spanish had been colonizing the coastal strip of California since the late 1700s, establishing missions built by Native people they had enslaved. In 1805 or 1806, Spanish soldiers were sent out from the Presidio, near the Golden Gate, to round up a group of Chupcan, or Bay Miwok, tribesmen who were not inclined to become enslaved at the Mission San José. The natives hid in a willow thicket near Pacheco, and the Spaniards camped nearby, intending to capture them in the morning. During the night, the Chupcan escaped north across the Sacramento River at the Carquinez Strait. The Spanish, finding no trace of their quarry, believed that the disappearance was only possible with the help of the Devil (el Diablo). The Spanish soldiers named the area Monte del Diablo, meaning "woodland [or thicket] of the devil." Whether or not this story is true, Monte del Diablo was a landmark on the Spanish map *Plano topográfico de la Misión de San José*, dated around 1824. In 1828, the name was used for the Rancho Monte del Diablo Spanish land grant given to Salvio Pacheco.

Mistaking *monte* for "mountain," American settlers later changed the name to Mount Diablo. In 1851, surveyors in the newly minted state of California were looking for a good landmark to use as the starting point for the state survey. They used Mount Diablo to establish the principal meridian and baseline for surveying the northern part of the state and all of Nevada. The point of reference for all maps in the region, the survey benchmark, is located in the visitor center at the summit.

View south to Mount Diablo from Suisun Bay (This image is taken from Sights in the Gold Region, and Scenes by the Way . . . With numerous illustrations . . . Second edition, 1850, revised and enlarged *by Theodore Johnson. The original is held and was digitized by the British Library.* https://commons.wikimedia.org/wiki/File:JOHNSON1850_ MONTE_DIABOLO_NEAR_SUISUN_BAY.jpg.*)*

STOP 14: MOUNT DIABLO

The rains haven't started yet this season, and I head south through rolling hills of golden-brown grass. Solitary oak trees dot the slopes, providing much-needed shade from the hot sun for the few cows that are grazing the hillside. Crossing the Carquinez Strait, where the combined Sacramento and San Joaquin Rivers empty into San Francisco Bay, the road leads past the cities of Concord and Walnut Creek, booming and vibrant suburbs of San Francisco. The foothills of Mount Diablo rise before me. I notice a lot of towering blue-green eucalyptus trees, with a slightly familiar, slightly medicinal herbal smell. This feels incongruous, out of place. Aren't these the trees that feed and house cute koala bears in Australia? What are they doing here?

As with so many good ideas gone wrong, the eucalyptus was introduced to California just north of here in the

Suisun Valley in 1853 as an ornamental tree. At the time, the Gold Rush was going strong, and thousands of people were moving to California. Wood was needed for homes, for firewood, for mine timbers—for just about everything—and the native oak and pine forests were being cut at a furious rate. Eucalyptus was known to be a fast-growing tree that thrived in a Mediterranean climate like that in California. It was thought that the wood was fireproof (it is, in fact, just the opposite), that the trees somehow strained impurities from the air, and that the oils had medicinal value. In 1872, Robert Stearns of the California Academy of Science wrote,

> When we consider the fact of the great number of farms in California that are nearly or wholly destitute of wood, and the great and continuous expense entailed by our system of fencing, the importance to the farmer of dedicating a portion of his land to the cultivation of forest trees, from which he can obtain fuel and fencing materials is too palpable to admit of debate....
>
> Of the *Eucalypti, H. globulus* is very common in California, and easily cultivated: it is the Blue Gum of Victoria and Tasmania. This tree is of extremely rapid growth and attains a height of 400 feet, furnishing a first-class wood; shipbuilders get keels of this timber 120 feet long; besides this they use it extensively for planking and many other parts of the ship, and it is considered to be generally superior to American Rock Elm.[1]

Soon eucalyptus trees were being planted all across the state. Today, 180 years later, they have become a nuisance: they grow like weeds, are extremely fire-prone (in a fire, the vaporized oils literally explode), and are invasive, crowding out native trees and sucking up groundwater, drying what little moisture is found in the soil. I leave them behind as I enter Mount Diablo's grassy foothills.

I have been up to the summit of Mount Diablo half a dozen times, and the sky over the Central Valley was never clear. What with coastal fog, forest fire smoke, and windblown dust swirling, at best I could make out (or imagine) thunderheads building over the Sierra some 85 miles off to the east. This time, as I drive up Mount Diablo, I am in a pea-soup fog and howling wind. It is now winter, and the grass is just beginning to sprout from the bone-dry fields. I can't see more than 30 feet through the gloom, and can barely find my campsite. The pine trees around my camp are catching the fog and raining on me. It is impossible to stay dry, but I am determined to stick it out because I have read that the views are magnificent after a good blow.

It gets dark by 6:30, and the night seems to go on forever as the wind gusts hurl fat drops onto the tent, which proceed to drip slowly but incessantly onto my sleeping bag. Finally, around midnight, the wind and drips stop. The sky is clearing. What had been a cool night becomes downright cold, and I snuggle deeper into my down bag. Eventually it grows light outside, and I go out. The sky is still dark enough that Venus shines bright, but the fog and clouds are gone. I drive the final few miles to the summit and am rewarded for my patience. To the west I can see the towers of downtown San Francisco over the East Bay Hills. To the east, beyond the flat expanse of the Central Valley, off on the horizon is a snowcapped Sierra Nevada. The views are worth the wait.

The summit of Mount Diablo has long been known for its expansive views. In 1862, William Brewer, of the nascent California Geological Survey, wrote:

> The summit was reached, and we spent two and a half hours there. The view was one never to be forgotten. It had nothing of grandeur in it, save the almost unlimited extent of the field of

view. The air was clear to the horizon on every side, and although the mountain is only 3,890 feet high, from the peculiar figure of the country probably but few views in North America are more extensive—certainly nothing in Europe.

To the west, thirty miles, lies San Francisco; we see out the Golden Gate, and a great expanse of the blue Pacific stretches beyond. The bay, with its fantastic outline, is all in sight, and the ridges beyond to the west and northwest. Mount St. Helena, fifty or sixty miles, is almost lost in the mountains that surround it.... South and southwest the view is less extensive, extending only fifty or sixty miles south, and to Mount Bache, seventy or eighty miles southwest.

The great features of the view lie to the east.... First, the great central valley of California, as level as the sea, stretches to the horizon both on the north and to the southeast. It lies beneath us in all its great expanse for near or quite three hundred miles of its length!... On the north are the Marysville Butters [sic], rising like black masses from the plain, over a hundred miles distant; while still beyond, rising in sharp clear outline against the sky, stand the snow-covered Lassen's Buttes, over two hundred miles in air line distant from us—the longest distance I have ever seen.

Rising from this great plain, and forming the horizon for three hundred miles in extent, possibly more, were the snowy crests of the Sierra Nevada. What a grand sight! The peaks of that mighty chain glittering in the purest white under the bright sun, their icy crests seeming a fitting helmet for their black and furrowed sides![2]

Brewer was off on the elevation of the peak by a whole 41 feet. Not a bad measurement in 1862, though.

The oldest rocks on Mount Diablo are Jurassic **ophiolites**. An ophiolite is a section of the Earth's oceanic crust and underlying upper mantle that has been uplifted and exposed at the surface. These ophiolites formed at the Pacific oceanic spreading center about 165 million years ago. Ophiolites typically consist of a sequence of rocks that, from bottom to top, include deep mantle rock (peridotite and serpentinite), vertical sheets of molten rock (**feeder dikes**, or **sheeted dikes**), and pillow basalts erupted onto the ocean floor.

The Mount Diablo ophiolite was caught in the subduction zone along the west coast of North America. It was dragged down as the oceanic plate plunged to depths around 12 miles, and then recently uplifted by thrust faulting. It is exposed on the northern flanks of the mountain.

The slightly younger Franciscan Complex is exposed at the summit of Mount Diablo and North Peak. The dark gray-green altered pillow basalt is interleaved with red-brown ribbon chert and graywacke sandstone. The basalt represents the ancient ocean floor; the cherts were deep-marine planktonic muck; and the graywacke contains sand over 100 million years old shed off the Sierra volcanic arc. There are a few exotic blocks mixed into the Franciscan Terrane, notably a 165-million-year-old **blueschist** (the gunmetal-blue color comes from the mineral **glaucophane**) that formed when ocean-floor basalt was altered by the high pressures encountered in a subduction zone. The blueschist is on display on the side of Summit Road about 280 feet beyond (above) the Rocky Point picnic area: it is the 6-foot-wide dark-blue boulder on the left when you're driving toward the summit. Right next to it is reddish ribbon chert.

The Great Valley Sequence was pushed up into nearly vertical ridges of rock all along the flanks of Mount Diablo. In this area, the sequence contains rocks that range in age from 5 to 140 million years old. The older sediments are derived from the Sierra: as mentioned at Monticello Dam, the strata were deposited in a marine setting mainly as turbidites. The younger sediments were contributed from the Diablo Range to the south and the Coast Ranges to the west. During this long period of time, the area at Mount Diablo was mostly below sea level, but sometimes it was uplifted above sea level and subject to erosion. We know this because there are few or no rocks of Paleocene, Oligocene, and early Miocene age.

On the way to the summit, it is well worth pausing at Rock City, an outcropping along Fossil Ridge (37.848402, −121.931814). Rock City, famous for the "wind caves" found there, is located in the Eocene-age Domengine Formation sandstone, deposited about 50 million years ago. The tan-colored sandstone is pockmarked with depressions, hollows, and dimples, like a rocky Swiss cheese, and also comes with abundant snail fossils of the genus *Turritella*, whose shells spiral upward like a miniature but really long soft-serve ice cream. The caves, as it happens, are not formed by wind. Rather, they were created by water seeping through fractures. The water dissolved the natural cement that holds the sand grains together, allowing the softened parts of the rock to be readily eroded. But when there is a good breeze blowing, listen for the wind whistling around and through the caves.

During the Eocene (about 56 million to about 34 million years ago), the area was a coastal plain. Plants grew in marshes, swamps, and lagoons beside a sandy beach. These plants were later converted to coal as a result of the heat and pressure of deep burial. The low-grade coal, or lignite, was mined out of the Domengine Formation north of Mount Diablo at the Black Diamond Mines. About 4 million tons were produced between the 1860s and 1906. At the time, it was the largest coal-mining operation in California. Guided tours of the coal mines are available at Black Diamond

Mines Regional Preserve: https://www.ebparks.org/parks/black-diamond.

Having spent much of my career as a petroleum geologist, and Mount Diablo being surrounded by oil country, it behooves me to say a few words about the history of oil and gas in California. Most Californians have a limited understanding of the saga of the energy industry in their home state. When they think of energy, most likely they think of the Diablo Canyon nuclear power plant that was built next to an earthquake fault, or they recall the 1969 Santa Barbara oil spill that reignited the environmental movement in the state. (A tone-deaf Fred Hartley, president of Union Oil at the time, famously denied that the event was a disaster: "I don't like to call it a disaster, because there has been no loss of human life. I am amazed at the publicity for the loss of a few birds.") But the truth is that California was literally built on oil: there is so much oil that it bubbles to the surface of its own accord. I saw this happening when I visited the La Brea Tar Pits in Los Angeles: oil seeps out of the grass in the surrounding park. I have seen it on the beaches of Southern California, where tar balls from natural offshore oil seeps float ashore to sully the feet of beachgoers.

California was blessed (or cursed, depending on your point of view) with abundant oil and gas because it was, for much of its geologic history, a series of shallow ocean basins rich in marine life. When abundant organic matter is buried rapidly and heated, it cooks to become oil and gas (indeed, petroleum geologists call such basins "hydrocarbon kitchens"). Since its discovery in California, oil, even more so than gold, has made the state an economic powerhouse. In 1865, seven years after Edwin Drake drilled the first oil well in Pennsylvania, the Union Mattole Company started producing oil in the Mattole Valley near Petrolia, in Northern California. By 1876, the center of the oil industry had shifted to Southern California. Los Angeles, in particular, owes its wealth and prominence to the oil fields first discovered in the 1870s and 1880s. By the 1920s, California was the nation's top oil-producing state. Even today it ranks seventh in oil production. To date, cumulative oil production in the state is well over three billion barrels.

Natural gas, like oil, is derived from buried organic matter. It is either a product of cooked woody material or is generated when oil is overcooked and breaks down into tar and methane. After they are generated, the buoyant oil and gas float upward through groundwater in porous rocks, usually sandstones, until they are trapped at the top of folds in the rocks. Those folds are the main targets of oil and gas exploration.

North and northeast of Mount Diablo, the Domengine Formation produces natural gas from large arch-shaped folds (**anticlines**) at Concord Field, Los Medanos Field, Willow Pass Field, and Mulligan Hill Field. The folds were formed by the same compression that uplifted Mount Diablo. That compression is a result of the westward movement of the North American Plate bumping into the Pacific Plate, kind of like pushing a rug against a wall until it rumples into folds.

Compression not only uplifted Mount Diablo but also tilted the rock layers around it. Adjacent to but younger than the Domengine, the fossil-rich Briones Formation was deposited between 24 and 10 million years ago. A dark-gray pebbly sandstone, this Miocene beach sand formed the Briones Formation sandstone. The tilted, near-vertical sandstone is the resistant layer that holds up Fossil Ridge on the lower western flank of Mount Diablo. The Briones Formation was used to build the visitor center at the summit. The best fossils were picked clean long ago at Fossil Ridge, although I have found some while exploring with my grandson. But it *is* a park, so no collecting or removing fossils. If there are none to be found here, look for abundant fossil clam, scallop, mussel, and oyster shells in the walls of the visitor center.

Nine million years ago, the Mount Diablo area rose above sea level for good. Fossil mammals, including mastodons, horses, camels, and rhinos, have been found in the Blackhawk Ranch quarry on the south side of the mountain.

The bulk of Mount Diablo is thought to have been uplifted in just the last 2.5 to 3 million years. Recent, ultraprecise satellite GPS measurements reveal that the mountain continues to rise between 1 and 2 inches every 10 years. The mountain is thought to have been formed by south- and southwest-directed thrusting, even though the Mount Diablo Thrust Fault is not exposed at the surface. South of Mount Diablo, the Greenville Fault, part of the San Andreas system, steps to the left to continue northward as the Concord Fault. The area in the gap between these faults is being squeezed in a vise, causing the Mount Diablo Thrust Fault to push up the mountain. The thrust fault puts Mount Diablo ophiolite (oldest) over the Franciscan Complex, and the Franciscan Complex is in turn thrust over the Great Valley Sequence (youngest). The mountain exists as a high point because the ophiolite is more resistant to erosion than the sediments of the Great Valley Sequence.

Continue driving up South Gate Road for 2.3 miles to the intersection with Summit Road and turn right. Take Summit Road to the top, another 4.5 miles, or 23 minutes total. Park at the Upper Summit parking lot (37.881713, −121.914673).

As I approach the summit, I am driving across progressively older layers. This is quite opposite the normal sequence one would expect—that is, the oldest layers on the bottom and youngest on top. What I see here is precisely a result of the thrusting of older rocks on top of younger rocks.

The summit museum has an excellent geology display. And, as I mentioned, the visitor center is built out of Briones Formation and is chock-full of fossil sea shells.

I walk for 1,000 feet or so along the Mary Bowerman Interpretive Trail. The trail starts at the Lower Summit parking lot (37.880861, −121.918176), about 1,000 feet west of the Upper Summit lot, and runs to an overlook, with fourteen stops along the way. A trail guide is available at the trailhead. Of particular interest are stops 3 (greenstone), 4 (graywacke), 5 (ribbon chert), and 6 (a contact between shale below and basalt above that suggests the rocks have been overturned). The quarry that is visible to the north produces crushed rock from the sheeted dikes portion of the ophiolite. Ophiolite, a hard and durable rock, is used in the building of roads and dams, as railroad ballast, and for building foundations.

Leaving Mount Diablo, I cross the Great Valley without stopping because there is little to stop for. Most of the surface is cultivated, and there are few rock outcroppings. On the other hand, the sheer variety of options on the car radio is enchanting: from northern Mexican polkas to conservative talk shows, from Johnny Cash to Taylor Swift and everything in between. Traversing the Great Valley is to cross a land of contrasts. There are giant, newly built mansions next to ramshackle homes with rusting tractors and trucks ornamenting their lawns. Billboards tell me that water is precious, yet I have seen sprinklers irrigating orchards at midday when the temperature hovers near 110°. Through these orchards there are roads as flat as a pancake and straight as an arrow, just inviting me to see how fast I can go, and cunning Highway Patrol cars waiting in ambush when my foot gets a bit too heavy on the gas pedal. I have been asked the dreaded question "Do you know how fast you were going?" and I don't want to repeat that experience. So I listen to the radio, keep my speed mostly under control, and head for the hills.

CHAPTER 3

Gold Rush Country

California's gold district lies in the western foothills of the Sierra Nevada between Downieville in the north and Mariposa in the south. A subset of this larger district, the "Mother Lode," is the zone of hard-rock gold deposits stretching north–northwest along the Sierra foothills. This zone, between 1 and 4 miles wide, extends 120 miles from Mormon Bar in the south to Georgetown in the north. Within that zone are hundreds of mines and prospects dating back to the Gold Rush. The gold is almost always found in white quartz **veins** that can reach 50 feet thick and extend several thousand feet. These gold-bearing quartz veins are what miners call **lode** deposits.

The Mother Lode, that famed source of gold, largely follows the Melones Fault zone, a roughly north–south geological boundary in the western Sierra Nevada foothills that is responsible for many fractures. These fractures were the conduits in which gold-bearing veins were emplaced.

The Melones Fault is a boundary between contrasting geologic **terranes**. East of the Melones Fault lies an older, metamorphic terrane, a landscape that has witnessed eons of transformation. The Paleozoic-age Shoofly and Calaveras Assemblages tell a story of geological burial and upheaval, of rocks that had been deposited on a continental margin, then altered and deformed by pressures and temperatures over great expanses of time. West of the fault zone is a younger,

Jurassic terrane. Here, metavolcanics, graywacke, and serpentinite tell the story of a time when molten rock flowed and ancient sediments settled in the depths of primordial oceans.

Gold is exceedingly rare on Earth. That's part of its allure. The ancient Egyptians and Greeks linked gold to the sun; the ancient Incas believed it was the breath of God, the sweat of the sun. Indeed, like all atoms heavier than hydrogen, helium, and lithium, gold was forged by the thermonuclear reactions taking place in the center of stars or when neutron stars collide. Atoms of the metal are then scattered across the universe when the star goes supernova. Gold in the Earth was brought by the asteroids that clumped together to form our planet and by interstellar gold dust falling to Earth. There is not much of it, and it is widely distributed in the Earth's crust.

For prospectors to find gold, it has to be concentrated somehow. Concentration usually takes place around the edge of magma chambers, those pools of molten rock lying deep underground, usually below volcanoes. As the molten rock pushes its way up to the surface, it creates fractures and **fissures** in the surrounding and overlying rocks. When fractures form, they are zones of lower pressure, and steam and hot fluids within the magma get sucked into the fractures, as when a pot on the stove boils over and lifts the top off the pot, and the lighter stuff spills all over. In magma, those fluids contain water vapor and silica, often iron, and sometimes rare metals such as gold. When they cool, the silica becomes quartz veins, the iron becomes **pyrite**, and the gold becomes, well, gold. The gold-quartz veins that would come to define the Mother Lode formed during the Early Cretaceous period, spanning from about 127 to about 108 million years ago. Like salt crystallizing out of your sweat, the veins in the California gold district crystallized out of hot magmatic fluids in fractures alongside the Melones Fault. Rather than the sweat of the sun, as the Incas believed, gold is more like the sweat of the Earth.

But gold in the Sierra foothills has two sources: in quartz veins and in river gravels. Gold is a unique element: it is inert; that is, it doesn't react with most of the other elements. It for sure doesn't react with oxygen, so it does not tarnish or rust. It is also very heavy, almost twice the density of lead. So when it does erode out of veins and move down a hillside and into a stream, it is always going to be shiny. And gold always settles into and concentrates where other minerals are winnowed away by currents or waves.

Like the ancient farmers who winnowed grain by tossing it into the air to let the wind carry away the chaff, the strongest stream currents carry away lighter minerals and leave behind gold. That is why gold is found in deep pools beneath waterfalls, below large rocks and boulders, among exposed tree roots, and along the inside bends of a stream. In fact, panning for gold does the same thing: it finishes the process of winnowing by taking advantage of the density of gold and washing all the lighter minerals out of the pan. Only the heavy gold flakes and nuggets are left behind.

Gold flakes and nuggets found in riverbeds are called **placer** gold. The zone of placer mining in California is longer and wider than the Mother Lode. That is because streams carried the gold away from its source. But that's not all. It's also because some of the placer gold was brought to California by ancient rivers that flowed out of Nevada, where other lode gold deposits exist. California placer deposits stretch almost 180 miles north to south, and the zone is almost 50 miles wide, extending from the floor of the Central Valley well up into the Sierra. This area is characterized by deep valleys and streams flowing off the mountain range. The deep valleys expose Jurassic and Cretaceous bedrock, and the high ridges contain remnants of the flat, low-lying landscape and river channels that existed before the Sierra Nevada was uplifted.

When I grew up in California, everyone learned all about the great Gold Rush in elementary school. My interest in looking for gold was one of the reasons I became a geologist.

I never found any in the wild, but there are still prospectors (professional and amateur) combing the hills for the elusive element, hoping to get rich; every year, people still find nuggets of the yellow metal. But there was a time when California was untouched by gold fever, when most of the state was populated by California Indians, and the Spanish along the coast got rich by running cattle in the coastal hills. That all changed in a big way in 1848.

THE SIERRA FOOTHILLS

The California Gold Rush is usually described in terms of intrepid prospectors, noble settlers, and the effect it had on the state of California. It was *the* seminal moment in the state's founding. I want to pause this geologic pageant and take a moment to describe the absolutely unmitigated disaster the Gold Rush was for the original Californians: the Indigenous people, the California Indians.

The impact of the Gold Rush on California Indian communities and their environment cannot be overstated. It was a catastrophe of the first order not only for the people but also for the landscape, native plant and animal communities, and bodies of water that still carry the scars and poisons of the period. For those Native people who had managed to survive the Spanish missions and the exploitation of private ranches, the Gold Rush was characterized by displacement and a cataclysmic loss of life caused by intentional, subsidized killing. The word *genocide* has been used loosely and way too often without much thought. What happened to the California Indians *was* a genocide.

The Yokuts were the most populous Indigenous group in California at the time of European contact. From about 14,000 in 1848, the population fell to 600 by 1880.

In 1846, the United States went to war with Mexico, and Americans in California rebelled against Mexico. Commodore Robert Stockton appointed John Frémont commander of the California Battalion and tasked him with monitoring the

California coast. John Sutter, a large landowner near the future site of Sacramento, needed soldiers to defend his fort and livestock, as well as to march on Southern California to make war on Mexican forces there. He persuaded Piupiu-maksmaks and Walla Walla men to join the California Battalion. He invited José Jesús and other Yokuts to join the war on the California Republic side. The Yokuts and Miwoks in Company H scouted for the battalion. Jesús led horse raids, leaving many of the Californios (Mexican landowners in California) riderless. In January 1847, the Californios surrendered at Cahuenga Pass, and Company H disbanded. After the war, Americans and Californios worried that the Native people would renew their horse raids.

In December 1846, General Stephen Kearny had arrived from New Mexico and appointed John Sutter, Mariano Vallejo, and Jesse Hunter as federal Indian agents tasked with ending the raids through diplomacy, gifts, or force. By March 1847, Indian raiding parties had swept along the coast. In December 1847, Yokuts raided ranches in the Livermore Valley and Mount Diablo area. Charles Weber, a partner of Sutter and founder of Stockton, reached out to Jesús to stop the raids. In February 1848, Jesús and two hundred Yokuts attacked a group of Locolumne raiders on the Calaveras River, killed some, and returned the stolen horses. Americans and Californios soon began preemptively attacking Indigenous communities. They targeted groups that had little to do with the raids, figuring the attacks would serve as a lesson to others.

In 1847, Andrew Kelsey and Charles Stone bought fifteen thousand head of cattle, twenty-five hundred horses, and the right to pasture them near Clear Lake, just north of The Geysers. They then forced Eastern Pomos and Clear Lake Wappos to work their ranch under the Mexican system whereby workers are transferred with the land. They imprisoned and whipped the Indians and raped the women. Eventually the Pomos had had enough and killed Kelsey and Stone with arrows. The US Army responded immediately.

On May 15, 1850, seventy-five soldiers arrived at what is now called Bloody Island at the north end of Clear Lake. Captain John Frisbie was directed to "exterminate" the Pomos and Wappos they held responsible. Whereas the Pomos came to negotiate, Frisbie ordered them shot. What followed was the largest massacre in US history, exceeding even Sand Creek and Wounded Knee: an estimated eight hundred Pomos were murdered.

When gold was found in the American River near Coloma, Indigenous people made up half of the four thousand miners by the middle of 1848. But most Americans believed that Indigenous people had no place in the United States. Between 1850 and 1851, California politicians created what historians have called a "killing machine." The governor authorized citizens to form militias to hunt down and kill Indigenous people. When state resources ran out, they asked the federal government to pay the expenses. In the 1850s and 1860s, the US government paid out over $1 million to fund these militias. Between 1846 and 1873, vigilantes, militias, the state, and the federal government initiated hundreds of campaigns that killed upwards of sixteen thousand California Indians.

Violence was endemic in mining areas. In early 1849, a group of miners from Oregon arrived at Coloma and attempted to rape a Nisenan woman. The Nisenans killed seven of the Oregonians. The surviving miners attacked a Nisenan town, killed four, took prisoners, and burned the town. Two days later they attacked another Nisenan town, killing thirty and taking forty to sixty prisoners. One Oregonian estimated that they killed "more than a hundred" Nisenans in a month.

In the summer of 1851, the Indigenous people along the Merced River told Indian agent Adam Johnston that they were starving and would die unless they returned to the hills. Silting from mining had clogged streams and ruined the salmon runs. Miners chopped down trees and cordoned off the land. The California Superintendency of Indian

Affairs was established in 1852 with the purpose of managing Indian affairs, enforcing policies, and maintaining the peace. By 1859, the federal government had abandoned most of the reservations, and the following year, Congress abolished the California superintendency. During the Civil War, the US cut funding for California reservations by 60 percent and reduced the number of reservations to three. The "Four Reservations Act" of 1864 reestablished the state superintendency, authorized no more than four reservations, abolished all previous acts that might conflict with the new law, and returned the previous reservations to the public domain to be sold to settlers. California settlers consistently called for ethnic cleansing, petitioning the state and federal government to forcibly relocate Indigenous people to reservations.

Between 1848 and 1860, California disenfranchised, disarmed, and legalized the indenture of Indigenous people. The US Congress failed to ratify treaties, reducing the ability of Indians to defend their interests. The US Army and state-authorized militias massacred Indians. The population of California Indians declined by about 80 percent, from 150,000 to 30,000.[1]

The Gold Rush towns of Grass Valley and Nevada City were built on the Nisenan towns of Daspah and Ustomah, respectively. The California Gold Rush was the beginning of the end for the Nisenan people. When the treaties of 1851 were not ratified by Congress, their fate was sealed. Nisenan lands were taken without compensation or replacement.[2]

The Gold Rush was a historic milestone for both California and California Indians. Humanity flooded into California, at first a trickle, but within the year a flood that inundated all the valleys, streams, and mountains from San Francisco to Yosemite and north. Native people were displaced by hundreds of thousands of newcomers. Their home territories were replaced overnight with bustling and lawless towns with no love for Indian people or culture.[3] As John Sutter put it in 1849, "The late emigrants across the mountains, and some

Oregon trappers and mountaineers, had commenced a war of extermination upon them, shooting them down like wolves, men, women, and children, wherever they could find them."[4]

The plight of California's Indians has largely been ignored in most histories of the California Gold Rush, at least partly because the settlers' concept of California excluded Indians. In the early years of American California, the California Supreme Court held that Roberto, an emancipated Indian, did not have the right to sell or transfer his land grant to a Californio because contracts involving Indians resembled those of "infants, idiots, lunatics, spendthrifts and married women." These were all citizens, but they required the "intervention of a tutor, curator, committee or guardian." The court upheld claims by American settlers that the land was in the public domain and thus open to settlement.[5] Thus California Indians were dispossessed of their land as well as their lives. It was not until the late 1800s and early 1900s that the federal government purchased lands for homeless Indians, creating a vast network of California rancherias. Then came the termination era. The California Rancheria Termination Acts were three acts of Congress and an amendment, passed in the 1950s and 1960s, that targeted rancherias for dissolution. Forty-six rancherias were successfully invalidated. Slowly, forty-one had their federal status restored.

These are among the long-term effects of the California Gold Rush. This is the legacy of the geology of gold and greed.

The classic American version of the Gold Rush story begins with a Swiss-German immigrant named John Sutter. In 1824, at the age of twenty-one, Sutter, a one-time apprentice printer and clothing store clerk, married the daughter of a rich widow in Burgdorf, Switzerland. Apparently, he was more interested in spending money than in earning it. Eventually, mounting debts brought charges that could have landed him in jail. So, like many others in his situation, he decided to leave his old life behind and head to America.

In May 1834, Sutter left his wife and five children behind

in Switzerland, and with a French passport, he traveled from Le Havre, France, to New York City, arriving in July 1834. Once in America, he reinvented himself as Captain John Augustus Sutter. From New York he moved to St. Louis, Santa Fe, Kansas City, Vancouver (Washington), Hawaii, and Alaska before heading to California.

When he arrived in 1839, Upper California was a province of Mexico. Sutter had to go to the capital at Monterey and get permission from the governor to settle in the territory. Governor Alvarado saw Sutter's plan to establish a colony in the Central Valley as a useful buttress to maintain the frontier against Indians, Russians, Americans, and British. In one of the greatest ironies of all time, Alvarado granted Sutter 48,400 acres to stop American encroachment into California.

In August 1839, Sutter began building a fortified settlement named New Helvetia, or New Switzerland, on the future site of Sacramento. In order to turn extensive forests into timber, he built a sawmill on the South Fork American River in the Sierra foothills near what is now Coloma.

On January 24, 1848, James Marshall, a carpenter working at Sutter's Mill, noticed shiny pebbles winking at him from the riverbed near the sawmill. As the ultimate proof that timing is everything, nary a week later, on February 2, Mexico signed the Treaty of Guadalupe Hidalgo, ending the two-year Mexican-American war. The treaty ceded large swaths of Nevada, Arizona, New Mexico, Utah, and Colorado, as well as California, to the United States.

Marshall brought the nuggets to Sutter, and together they did some primitive tests to confirm whether they were actually gold. Sutter concluded that they were, but he didn't want the discovery to disrupt his plans for building a settlement and farming. Nonetheless, just to be safe, he acquired title to as much land near the discovery as possible.

Much as Sutter tried to keep the gold discovery quiet, he failed fabulously when a merchant and newspaper publisher,

Sam Brannan, returned to San Francisco from Sutter's Mill and began to publicize the story. News of the discovery of gold spread like wildfire and began the California Gold Rush of 1849. By 1856, over three hundred thousand prospectors and pioneers, known collectively as "forty-niners," had arrived in California to look for the elusive metal.

The original forty-niners panned for gold nuggets and flakes in river-channel gravel and sand. When the gold in the rivers and streams had pretty much run out, the miners went looking for the source, for where the gold had come from. Some gold had come from white quartz veins poking out of the ground and slicing across rocky outcrops. Other gold had washed down the hillsides from ancient riverbeds now exposed high on the valley walls. Once they determined that these old streambeds also had gold, the prospectors became desperate to get at it.

By 1853, some clever miners had come up with a method for doing so. Water was diverted from streams high above the area to be mined; it was channeled, sometimes for many miles, in ditches, flumes, and hoses, to large water cannons sitting at the base of a hillside. The water, now under high pressure, was shot by the cannons as jets of water up against the hillsides. Everything on the hillsides, including the ancient streambed gravel, trees, and soil, was washed down to where miners could get at it. They used giant **sluice boxes** to sift through the sediment for the rare gold nuggets. Hillsides became barren, deeply eroded wastelands of white to rusty-red slashes across the dark, tree-covered hills. These became known as "the diggings," and the technique as hydraulic mining.

This method of mining spread quickly in the foothills north of Sacramento. Along a nearly 60-mile stretch, the slopes became denuded. With no trees to slow the erosion, the loss of soil accelerated. So much sediment was washed downstream in the 1860s that Yuba City and farms in the Sacramento Valley were buried in a muddy slurry. Sacramento was flooded repeatedly, and San Francisco Bay turned a muddy brown. San Francisco Bay, already shallow, was filling with mud at the rate of 1 foot per year. It looked as though nothing could stop it.

In one of the first-ever environmental actions, farmers devastated by the mud and repeated flooding formed the Anti-Debris Association and filed a lawsuit against North Bloomfield Mining and Gravel Company in 1883. In 1884, Judge Lorenzo Sawyer found that the hydraulic operations were harming those downriver and banned all discharge of mud and waste into rivers. After thirty years of environmental devastation, the Sawyer Act effectively ended hydraulic mining in California.

Since the first nugget was found, more than 750,000 pounds of gold were taken from the American River *alone*. And this was not only flakes and pebble-size nuggets. The largest "nuggets" found here include the 54-pound Willard nugget, found in 1859 at Magalia; the 103-pound Monumental nugget found at the Sierra Buttes mine in 1869; and the 195-pound Carson Hill nugget, found at the Morgan mine in 1854.

High-pressure water jets from water cannons at the Malakoff Diggins, 1850s or 1860s. Eroded material is washed down to sluices that winnow out the gold. This was immensely destructive to the landscape. (https://commons.wikimedia.org/wiki/File:North_Bloomfield_Mine_(Malakoff_Diggins)_-_a7195.gif)

These were huge at the time, when gold was selling for $20 per ounce. A 54-pound nugget would have been worth $17,280 at a time when a new Henry rifle cost $40–$50 and a fancy seven-shot Sharps rifle cost $50. A good riding horse would run you around $75. Think what it would be worth today, when gold is hovering around $2,300 per ounce.

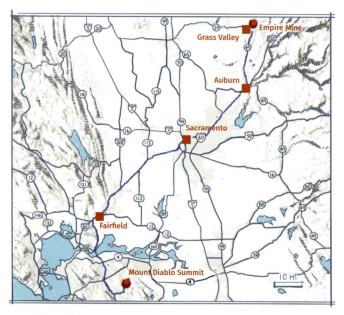

Map of the route from Mount Diablo to the Empire Mine

It is still possible to pan for gold in the rivers and streams of the Sierra foothills. According to KCRA-TV in Sacramento, several years ago a modern-day prospector found a 6-pound nugget in Butte County just north of Sacramento. The combination of forest fires and torrential rains during 2022–23 washed new gold from the hillsides and started a mini gold rush. But how do people know what to look for, and that they found gold? Years ago I worked for a mining company near Warren, in the wilds of central Idaho. I was having a beer with an old-timer after a hard day of collecting samples, and asked him that very question:

How do I know if I've found gold? He winked as if to let me in on a secret, and with a twinkle in his eye and a gravelly voice he said, "Kid, nothing *smiles* at you like gold."

Prospectors also found gold in veins. The next stop is at one of the biggest lode gold mines in the California gold belt.

To get from Mount Diablo summit to the Empire Mine, backtrack down the mountain and get on I-680 going north; in Cordelia, get on I-80 going east and follow it past Sacramento to Auburn; at Exit 119B, get on CA-49 going north to Grass Valley; take CA-20 West/Empire Street east and follow the signs to Empire Mine State Historic Park. Pull into the Empire Mine parking area (39.207169, −121.045678), for a total of 148 miles (3 hrs 10 min).

STOP 15: EMPIRE MINE STATE HISTORIC PARK

Empire Mine State Historic Park is on the National Register of Historic Places; it is also a federal historic district and a California Historical Landmark. The park showcases the Empire Mine, which operated from the mine's discovery in 1850 through its close in 1957. During that time, the mine produced 5.8 million ounces of gold, enough to fill a box 7 feet long by 7 feet high and 7 feet deep. It is described as "the oldest, largest, deepest, longest, and richest gold mine in California."

As I have said before, all metals are originally dissolved in magma deep in the Earth. As the magma cools, some minerals crystallize early and others late. Gold is part of the late-crystallizing magma and, along with silica and water vapor, is injected under high pressure into fractures in the surrounding rock. The precious metals and silica cool to form gold-silver-quartz veins, or lode gold deposits.

Gold ore at the Empire Mine was found in white quartz veins cutting across the granodiorite that makes up the bright foundation rock of the Sierra Nevada. The quartz veins carried 3–7 ounces of gold per ton of rock. Thus the miners had

Empire Mine yard

to excavate a ton of ore to yield, at most, 7 ounces of gold. At the time, gold was worth a bit over $20 an ounce. Working with a pickaxe, sledgehammer, and black powder (dynamite wasn't invented until 1867), miners still found it highly profitable to mine a ton of rock for a return as little as $60.

The first successful prospector in this area was George McKnight, who found gold in the Ophir quartz vein near Grass Valley in 1850. This was the beginning of the Empire Mine. Since this was "hard-rock" or lode mining rather than placer mining, there was a need for experienced hard-rock miners. The mine owners decided to bring over some of the best hard-rock miners in the world, Cornish tin miners from England, starting in the 1870s. The vein was mined by digging and blasting inclined shafts that followed the vein downward. As the miners burrowed into the ground, they invariably encountered the water table, usually pretty near the surface. The Empire Mine had to pump up to 18,000 gallons of water a day to keep from flooding. Eventually the increasing cost of mining at great depths, ventilating the tunnels, and pumping water made the mine unprofitable. It finally closed in 1957. But not before the deepest tunnels had reached depths of 9,015 feet. The mine entrance is around

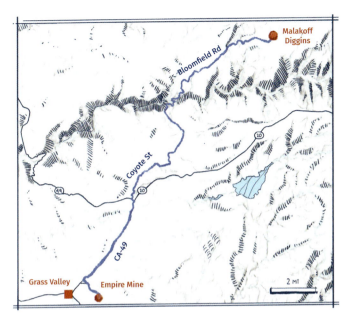

Map of the route from the Empire Mine to Malakoff Diggins

2,500 feet elevation, so the deepest tunnels reached more than 6,500 feet below sea level.

It is a little hard to picture, but this mine has 367 *miles* of abandoned and mostly flooded mine tunnels and underground workings. Even though the mine is only open to the 40-foot level, it is quite impressive to consider that this was originally chiseled out of the rock with fairly primitive tools. The mine yard has extensive displays of old mining equipment, with demonstrations and tours, a machine shop and blacksmith shop, and a museum in the 1890s-era owner's mansion. The museum has a complete model of the underground workings, and a mineral and ore collection.

The next two stops are in the heart of hydraulic mining country. The first stop looks at what's left of the wasted hillsides, now mostly overgrown. The second is a roadcut through the ancient river gravels that brought gold from western Nevada before the modern Sierra Nevada existed as a mountain range.

To get to Malakoff Diggins, return to CA-20/CA-49 (Golden Chain Hwy) and drive northeast to Exit 186; turn left (west) on Boulder Street, drive over the freeway, and make an immediate right (north) onto Coyote Street; take Coyote Street to North Bloomfield Road (closed in winter); turn right (north) onto North Bloomfield Road and drive to Malakoff Diggins, 21098 Relief Hill Road (39.368376, −120.899497), for a total of about 20 miles (50 min).

The Diggins overlook and trail are about 1 mile before reaching the North Bloomfield townsite. The last stretch before reaching North Bloomfield is a 5.5-mile section of unpaved road with washboards, rocks, and potholes. High-clearance vehicles are recommended but not required. This road also contains steep, narrow stretches.

STOP 16: MALAKOFF DIGGINS STATE HISTORIC PARK

The spoil heaps and scarred slopes of 150 years ago have largely returned to a natural state as the forest has reclaimed the desolation left by the old-timers. But destruction and healing are asymmetric in time. The damage occurred in just a handful of years; the recovery has taken a century and a half. There is a lesson in this for all of us.

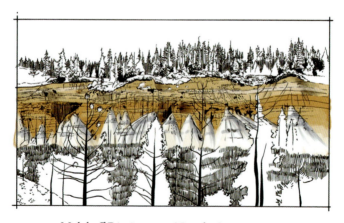

Malakoff Diggins, near Nevada City, 140 years after hydraulic mining ended

The old hydraulic **badlands** are mostly grown over with trees now, although I can still see where light-colored rock was exposed by the hydraulic miners. Barren, serrated **escarpments** 200 feet high mark where giant water cannons blasted away soil and rock to uncover gleaming bits of yellow metal. A walk through the forest reveals mounds of gravel and piles of large cobbles left behind by the miners, which are slowly being overgrown by bush and tree. Every now and then I come across a bit of historical litter, fragments of oversized iron or wrecked machinery that are slowly being reclaimed by nature. It is so quiet now that it is not easy to imagine the energy from a time when thousands lived and toiled here. I find myself peeking into the old wooden structures and running my hands over an old monitor—a rusted water cannon—to try to get a feel for it.

Whereas the Empire Mine was the embodiment of hard-rock mining, Malakoff Diggins State Historic Park

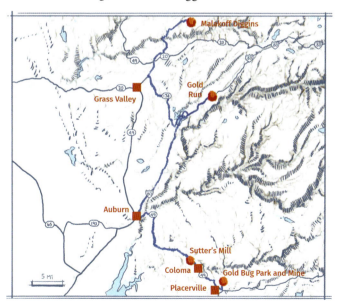

Map of the route from Malakoff Diggins to Gold Run, Sutter's Mill, and Gold Bug Park and Mine, Placerville

memorializes the archetype of hydraulic mining. In fact, it was the largest hydraulic mining operation in the state, over 7,000 feet long and 3,000 feet wide. Up to 600 feet of hillside was washed away during the mining operations. The town of Humbug grew up around one of the original operations on Humbug Creek. The town's name was later changed to the more respectable North Bloomfield. Today it is pretty much a ghost town populated by the occasional transient tourist.

I check out the visitor center where there is an exhibit that re-creates mining-era life. Several buildings dating back to the 1850s have been restored or rebuilt. Exhibits explain the process of hydraulic mining, and self-guided tours through the Gold Rush town are available.

To get from Malakoff Diggins to Gold Run, return southwest on North Bloomfield Road to Coyote Street; turn south on Coyote Street and drive to CA-49; turn left (east) onto CA-49 and drive to Exit 183; turn left (south) on Brunswick Road/CA-174; turn left (east) onto Rollins Lake Road and continue to the I-80 East on-ramp at Magra. Take I-80 east to Exit 145 just east of Gold Run. Take the Ridge Road turnoff and circle back to westbound I-80. Slowly drive by the roadcut (39.188814, −120.838003), for a total of 36.7 miles (1 hr 11 min). The narrow shoulder means there are no safe pullouts here.

As I drive east up Interstate 80 and look south, there are places where the forest opens up, and I can see remnants of the flat pre-uplift surface. That old plain has been tilted maybe 15 degrees or so to the west. I took the train once, the California Zephyr, from the Bay to Reno. From certain angles, as the train climbed the western slope of the Sierra, I could see not only the inclined tableland but also the deep canyon of the American River carved into that ramp. There are advantages to taking a train to see the country: you can focus all your attention on the landscape and at the same time enjoy a fine meal—though it's a lot harder to stop and see the rocks.

Watch for outcrops of Jurassic metamorphic **country rock** while driving to Gold Run, especially along Rollins Lake Road. The metamorphic rocks are recognizable by the thin, wavy layering, or schistosity, that indicates deformation under intense pressure.

STOP 17: PRE–SIERRA NEVADA STREAM GRAVELS, I-80 ROADCUT AT GOLD RUN

Fifty million years ago, Nevada didn't look like it does today. There was no desert, no Basin-and-Range of alternating valleys and mountains. It was a lush, temperate highland with dark forests, large lakes, and broad rivers. The dinosaurs had died out 16 million years earlier, and strange-looking mammals, the miniature ancestors of the horse and camel, roamed the valleys and hills. Farther east, in what would become the Rocky Mountains of Utah and Wyoming, compression caused mountains to rise as the North American Plate overrode the Farallon Plate. In California there was no Sierra Nevada, just a lowland consisting of the eroded remnants of the Sierran volcanic arc. The rivers that drained Nevada flowed west across this lowland to a coastline that shifted back and forth across what is now the Central Valley. These river channels were filled with sand, gravel, and cobbles that had eroded off the highlands in Nevada to the east. They also carried gold.

Jump forward 40 million years, give or take, and the modern Sierra Nevada began its rise. Around 10 million years ago, during the Miocene, a series of large north–south faults developed along the west side of the Owens Valley. This was the East Sierra Fault system. The west side moved up while the east side dropped downward. This caused a large block of the crust to slowly rise and tilt westward. Today, there is over 15,000 feet of vertical separation between granite at the top of the Sierra and granite beneath the Owens Valley to the east. This estimate of the amount of uplift represents a minimum, since much granite has been eroded off the highest peaks.

Some of the rivers, like the ancestral Yuba River, were able to maintain their level, at least for a while, by eroding down through the uplifting block. Other river channels were abandoned, left high and dry by the rising range. Eventually all the rivers flowing west out of Nevada were cut off, and the bulk of the state became an internally drained basin with no outlet to the sea. The emerging Sierra developed a new set of rivers and streams that cut into the mountains, forming the valleys we are familiar with. The new streams exposed the gold veins of the Mother Lode, contributing gold to the placer deposits in the modern rivers. Meanwhile, moisture-laden air, rising over the new mountains, cooled and dropped its precious load of rain and snow on the Sierra Nevada, an appropriate name, in that it means Snowy Sawtooth Mountains in Spanish. The areas to the east were left in the rain shadow of the range.

In 1851, miners following the gold upstream encountered ancient gold-bearing gravel, the so-called **auriferous gravels**, at Dutch Flat, an area of rolling hills 3,100 feet above sea level in the western Sierra. It soon became the major mining camp in the area, with over six thousand residents in 1853. In 1854, at

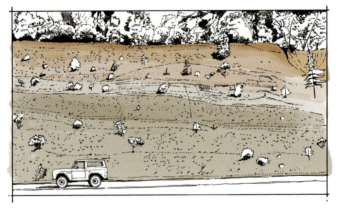

I-80 roadcut at Gold Run. These massive riverbed gravels came from Nevada. The roadcut extends nearly 2,500 feet.

the height of the Gold Rush, O. W. Hollenbeck founded the town of Mountain Springs a few miles southwest of Dutch Flat. It had a post office and became known for hydraulic mining of the old river gravels. In 1863, the post office moved one mile north and the town's name was changed to Gold Run. An estimated $6,125,000 in gold (at $20 per ounce) was mined in the period 1865–1878 alone. It was one of the richest gold-mining areas in the state. It all came crashing down when the Sawyer Act banned hydraulic mining and the discharge of sediment into rivers.

I want to see what these gold-bearing gravels look like, and there is a place where they are spectacularly displayed. The roadcut at Gold Run exposes 30–40 feet of sand, gravel, and cobbles deposited in ancient river channels that flowed from Nevada across the present-day Sierra Nevada. These roughly 50-million-year-old fossil Yuba River gravels were mined for gold here, at the Malakoff Diggins, and throughout the foothills gold district. Hydraulic mining operations occurred throughout the gold district wherever fossil stream channels were found. There are few locations where the gravels are exposed this well.

To get from Gold Run to Sutter's Mill, drive west on I-80 to Auburn; take Exit 119C to Elm Ave and head east to CA-49; turn left on CA-49 and drive south to Marshall Gold Discovery State Historic Park, 310 Black Street, Coloma (38.804504, −120.894003), for a total of 41.1 miles (52 min).

STOP 18: SUTTER'S MILL AND MARSHALL GOLD DISCOVERY STATE HISTORIC PARK, COLOMA

This is where it all began. In 1847, here in the Sierra Nevada foothills, nestled along the banks of the South Fork American River, lay a place untouched by the frenzied footsteps of gold seekers who would soon flood into the region. Indeed, the name comes from the Nisenan Indian name for the valley in which it is located: Cullumah, "beautiful." Then it was over: a sparkle in the bed of the river at Sutter's lumber mill caught the eye of James Marshall, and the great California Gold Rush was on.

As the news broke, waves of gold seekers descended on the region like locusts. The tranquil landscape was transformed into a furious hub of activity. Tents sprang up; makeshift markets, hotels, and saloons were built; and the riverbanks echoed with the hopeful, chaotic babble of men from all corners of the Earth.

The once-peaceful valley became unrecognizable. During the peak of the Gold Rush, from 1849 to 1860, the town's population rose to 888. And while the discovery of gold brought prosperity to many, the relentless drive for fortune also brought disputes, crime, and failure. Many left the gold fields as poor as they arrived.

As the years passed, the town of Coloma evolved, adapting to the ebb and flow of miners. Sutter's Mill and the town around it, once a symbol of industry and innovation, are now monuments to a brief moment of discovery and exuberance in what is effectively a ghost town.

The original gold discovery site on the American River, along with a reconstruction of Sutter's Mill, became part of California's state park system in 1927. Today, visitors can walk the paths of history, stand on the banks of the American River, and imagine the chaos and dreams that marked the founding of Coloma—a tribute to the irresistible allure of gold. Because really, the gravel bars along the river don't look particularly special.

Coloma sits in the foothills of the majestic Sierra Nevada. The volcanic arc that once spewed fire and ash has long since gone silent; all that is left is an uplifted block of granite, the igneous **basement rock**. Carried in the magma beneath the volcanoes, the metals that crystallized in veins as the magma solidified or that were carried in **hydrothermal fluids**—naturally heated water rich in minerals and gases—are mostly mined out. The gold-bearing quartz veins were subjected to

Photo by R. H. Vance shows Sutter's sawmill, on the American River near Coloma, circa 1850 (https://en.wikipedia.org/wiki/Sutter%27s_Mill)

erosion and the endless flow of water that carried the vein gold into the streams that wind through the foothills. There the heavy gold flakes and nuggets were concentrated into placer deposits by the winnowing action of flowing water. Over time, the placers grew heavy with gleaming fragments, just waiting to be discovered.

Standing here, it is easy to get carried away by all the history that has marked this place. Men got rich, got poor, and got killed for a handful of gold dust. Dreams were made and lost. People came and went. Those who remain, many of them of Swiss-Italian descent, are more into farming and ranching. Today's gold is in the ranches, vineyards, and orchards that dot the countryside.

Marshall Gold Discovery State Historic Park has a museum and many of the original and restored buildings. I buy a gold pan at the museum shop and learn how to pan for gold. My hands get really cold while panning in the frigid streams. I must not have the technique down, because I don't find any flakes.

There are video presentations in the museum, and sawmill demonstrations at the working replica of Sutter's original mill. There are self-guided tours of the town, the cemetery, the Monument Trail, and the Monroe Ridge Trail.

When I ask about the Gold Rush history of Indigenous people, the response from Gold Discovery Park is, "We don't have any information about Native Americans." The park has an exhibit that features Native people before the arrival of prospectors, but not during the Gold Rush. The tours describe the miners' existence; details about Native life during that time are sparse. There are not that many resources that describe Indigenous life during that period. The park does have an exhibit with Indigenous artifacts. Marshall Gold Discovery State Historic Park's "Living History Days" feature docents dressed in period clothing, who discuss the 1850s and give demonstrations in rope making, candle dipping, Dutch oven cooking, sawmill wood working, and games.[6]

PLACERVILLE

The Gold Rush town of Placerville was at one time the beating heart of California's Mother Lode. A mining camp was established there in the spring of 1848, and soon the prospectors realized that at least some of the gold in the creeks was coming from old river gravels high above the camp. They tunneled into these Eocene gravels, known as "dry diggings," and named the camp Old Dry Diggins. These dry placers were excavated both by using hydraulic mining and by tunneling into the gravels of Texas Hill. While scouring the hillsides, the prospectors also found vein gold deposits.

The town's name was changed to Hangtown after three men were convicted of robbing a successful miner in his hotel room in 1849. They were charged with robbery and murder and sentenced to death by hanging. A large oak tree on Main Street was chosen for the triple hanging. Sometime later, the tree was cut down and the stump ended up in the basement of a saloon. The building has been restored several times, and now houses Hangman's Tree Ice Cream Saloon at 305 Main Street.

The town quickly became a supply center for the nearby mining camps. The Methodist Episcopal Church was built of brick in 1851 and still stands, making it the oldest continuously used church building in El Dorado County. Newspaper ads for the stage coach running through town began using the name Placerville in 1852. Perhaps in an effort to make the town more family-friendly, it was officially incorporated as Placerville in 1854, although the creek wending its way next to Main Street is still Hangtown Creek.

Fires destroyed much of the town in 1854 and again in 1856. But there's no keeping a good town down. Placerville became the county seat for El Dorado County in 1857. A stagecoach line crossed the Sierra Nevada along the Placerville-Carson Trail, established the same year. In 1859, silver was discovered in Nevada at the Comstock Lode. Suddenly, prospectors and miners left for Nevada. Much of the silver traveled back west through Placerville on its way to becoming silver dollars at the San Francisco mint. Placerville became a relay station for the Central Overland Pony Express mail, the 1860 version of high-speed communication. The daring and heroic but short-lived Pony Express was abandoned in 1861 as telegraph lines stretched across the West.

By 1888, Placerville had two thousand inhabitants and a new railway connection to Sacramento. Celebrating the arrival of the railroad, one speaker called the day the union of the King of the Valley (Sacramento) with the Queen of the Mountains (Placerville). Unfortunately, fire struck Placerville again in 1891, burning many downtown businesses to the ground.

A torrential downpour during January 1896 washed new gold out of the hills and into the streams, starting a fresh gold

Hangtown, 1849. Originally named Old Dry Diggins, the name was changed to Hangtown and later Placerville. (Unknown photographer, UC Berkeley Library, https://digicoll.lib.berkeley .edu/record/8587#?xywh=-840%2C-65%2C3179%2C1292)

rush. People were finding fine and coarse gold in wagon ruts in roads and road ditches. Then, in 1911, another rich vein of gold was discovered within the town. The ore from "the gold bearing ledge" on the Stevens property assayed at $50 per ton. To put that in perspective, a ton of rock had to be processed to get 2½ ounces of gold. And that was good ore.

In addition to its gold, Placerville has two other claims to fame. A gentleman by the name of John M. Studebaker spent a few years working in Hangtown in the early Gold Rush years. He repaired wagon wheels and built wheelbarrows for miners. In 1852, "Wheelbarrow Johnny" took his savings and moved back to Indiana. There he started building wagons and established the Studebaker Wagon Corporation. Late in his career, he began making the automobiles that carried his name. (The last ones were produced in 1966.) The local boy who made good is honored in Placerville each year at the county fair, where the Kiwanis Club sponsors the Studebaker Wheelbarrow Races. Competitors navigate a muddy obstacle course with a vintage wheelbarrow, a mining pan, and a sack of ore.

According to local legend, in 1849 a miner struck gold nearby and headed into Hangtown for a night on the town. At the El Dorado Hotel, he decided to celebrate his newfound status as a rich man by ordering a lavish meal with the most expensive items on the menu. The cook fried up an omelet dish with all three of the most costly ingredients: bacon and eggs and oysters. Thus was born the "Hangtown Fry."[7] Apparently, it is still a Thing, because you can still find it on menus in town and near San Francisco.

To get from Marshall Gold Discovery State Historic Park to Gold Bug Park and Mine, drive south on CA-49 to Spring Street in Coloma; turn left (east) and drive to Pleasant Street; turn left (north) on Pleasant Street and drive to Bedford Ave; turn left (north) on Bedford and drive to Gold Bug Lane; turn right (east) on Gold Bug Lane and drive to 2635 Gold Bug Lane, Placerville (38.743343, −120.799347), for a total of 9.6 miles (20 min).

STOP 19: GOLD BUG PARK AND MINE

The historic Gold Bug Mine was originally called the Hattie Mine after the eldest daughter of William Craddock, one of the owners. Located in the heart of California Gold Rush country, the vein was probably staked by miners as early as 1848, but was not in production until 1888. Prospectors following the placer gold upstream from Sutter's Mill soon arrived at Big Canyon Creek, only 9 miles away. As the streambed gold eventually played out, they began looking upslope for the source of the gold. Mostly they were looking for **"bull quartz,"** the massive white quartz veins that sometimes also carried gold. And they found it. The gold-bearing veins, part of the Mother Lode, were found cutting across the **phyllite** bedrock, a metamorphosed mudstone. The forty-niners started by picking at the surface veins, then followed them underground, drilling and blasting ever deeper in their quest for more and richer deposits. The large chunks of broken rock were taken to a **stamp mill** where they were crushed into a fine powder, all to make the final extraction of gold easier.

The mine's gold vein was first excavated on the north side of the valley. As this vein played out, a larger extension was found on the south side of the valley. The Priest Mine worked an upper extension of the same vein.

In 1902, the claim was sold to Thomas Bishop and Frank Monaghan. It was sold again in 1926, when the new owner, John McKay, named the workings the Gold Bug Mine. McKay laid tracks for the ore carts and put in the stamp mills to crush the ore. An air shaft was drilled to provide fresh air to the miners. The mine continued operations until 1942, when President Roosevelt ordered all gold mines closed because they were deemed nonessential to the war effort, and men were desperately needed elsewhere. It is not known how much gold was recovered from the mine, as no records were kept.

The last private owner of the mining claims was William Meagher of Oakland, California, who used the land as a vacation retreat. Because the mining claims were no longer worked, the Bureau of Land Management took over the land as public property in the 1950s. The federal government offered to lease the property for recreational use, and in 1965 the City of Placerville took the lease with the promise that the land would be used only for recreation. In April 1980, Hangtown's Gold Bug Park Development Committee was formed to clean up the property for public use.

On February 1, 1985, the park was approved for listing in the National Register of Historic Places and also as a California Point of Historical Interest. Tours and walks around the grounds provide insights into the mining techniques and conditions of the Gold Rush era. The 61-acre park, with a stamp mill, prospect holes and trenches, and underground workings, provides visitors with a glimpse into the challenges and accomplishments of the miners who played a vital role in shaping the history of California during this period.

Placerville is the only city in California to own a gold mine. The park offers gold mine tours, outdoor exhibits of restored mining equipment, a working blacksmith shop, and gold- and gem-panning activities. Tours through the 352-foot underground mine tunnel are available. It is typical of the small-lode mines of the era. Find more information at the park's website: www.goldbugpark.org.

To get from Gold Bug Park and Mine to the El Dorado County Historical Museum, return to Spring Street and drive southwest to US-50 West; merge onto US-50 and drive to Exit 44B, Placerville Drive; turn right (north) on Placerville Drive and drive to 104 Placerville Drive (38.724534, −120.834589), for a total of 3.5 miles (8 min).

STOP 20: EL DORADO COUNTY HISTORICAL MUSEUM

The El Dorado County Historical Museum, near historic downtown Placerville, serves as an introduction and gateway to the rich history of the region, particularly its ties to the California Gold Rush. Key exhibits display Gold Rush and local history. The museum showcases artifacts, documents, and historical photographs, and provides tours related to the California Gold Rush, which had such a profound impact on Placerville and the surrounding area. There is an old general store and mining equipment, with wagons and railroad artifacts in the museum's yard, including the Diamond & Caldor's narrow-gauge Shay Locomotive no. 4.

Visitors learn about the miners, the techniques they used, and the broader social and economic consequences of the Gold Rush era. The museum delves into the history of El Dorado County, covering such topics as the Native Americans in the region, the early settlers, the impact of transportation (stage coach routes, the Pony Express), and the growth of the county over the years.

Special exhibits on various topics keep the museum's offerings fresh and appealing. Get the latest information from its website: https://www.edcgov.us/Government/Museum.

CHAPTER 4

Sierra Nevada

The Grand Canyon. Yellowstone. Yosemite. These are iconic national parks. Yosemite Valley is carved into granite of the Sierra Nevada, a range described by John Muir as "pervaded with divine light." This was the birthplace of the environmental movement in the US.

This magnificent valley was appreciated long before modern environmentalists championed it. Yosemite Valley has been inhabited by humans for at least the last 10,000 years. The Native people who lived there call themselves Ahwahnechee, meaning "dwellers in a deep grassy valley." The Ahwahnechee people include parts of the Miwok, Paiute, and Mono tribes.

The first Americans came into the valley during the Gold Rush. Those who had lived in the valley for generations didn't like or want any newcomers. In 1851, in an attempt to suppress Native American resistance, the US Army under Major Jim Savage led the Mariposa Battalion into Yosemite Valley: they were chasing about two hundred Ahwahnechee led by Chief Tenaya. The army eventually captured Tenaya and his band and burned their village. Letters home from members of the battalion helped to popularize the wonders of the valley.

One of those attracted to settle in the valley, Galen Clark, came across and publicized giant trees at the Mariposa Grove in 1857. In 1879, the Wawona Hotel was built for tourists visiting the giant sequoia trees. One of them, the Wawona

Tree, or Tunnel Tree, was 227 feet tall and 90 feet around. A carriage tunnel was cut through the tree in 1881 and immediately became a popular tourist attraction. The tree was estimated to be 2,300 years old. I could be wrong, but I don't think this treatment of a historic tree would be acceptable these days.

Concerned about commercial exploitation, prominent citizens advocated to protect the valley. The Yosemite Act, permanently closing the area to settlement, was signed by President Lincoln in 1864. It created the Yosemite Grant and entrusted it to the state of California. This was the first case of parkland being set aside specifically for preservation by the US government, and it set a precedent for the creation of Yellowstone as the first national park eight years later, in 1872.

John Muir, a Scottish-American sheepherder, naturalist, and explorer, wrote articles popularizing the area and increasing scientific interest in it. Muir was the strongest advocate of preserving the Sierra Nevada, having hiked and climbed every corner of Yosemite. His eloquence and powers of observation come across in his 1911 journal, *My First Summer in the Sierra*:

> August 26. Frost this morning; all the meadow grass and some of the pine needles sparkling with irised crystals, — flowers of light. Large picturesque clouds, craggy like rocks, are piled on Mount Dana, reddish in color like the mountain itself; the sky for a few degrees around the horizon is pale purple, into which the pines dip their spires with fine effect.
>
> The day has been extra cloudy, though bright on the whole, for the clouds were brighter than common.... Probably more free sunshine falls on this majestic range than on any other in the world I've ever seen or heard of. It has the

brightest weather, brightest glacier-polished rocks, the greatest abundance of irised spray from its glorious waterfalls, the brightest forests of silver firs and silver pines, more star-shine, moonshine, and perhaps more crystal-shine than any other mountain chain, and its countless mirror lakes, having more light poured into them, glow and spangle most. And how glorious the shining after the short summer showers and after frosty nights when the morning sunbeams are pouring through the crystals on the grass and pine needles, and how ineffably spiritually fine is the morning-glow on the mountain-tops and the alpenglow of evening. Well may the Sierra be named, not the Snowy Range, but the Range of Light.[1]

Muir saw how the coastal redwoods and old-growth spruce along the west coast were being cut for lumber, and knew that the natural wonders of the Sierra were both delicate and vulnerable. This made preservation of wildness a mission, an urgent undertaking for him, and his drive to conserve nature was instrumental in the creation of the Sierra Club.

Overgrazing of meadows by sheep and logging of giant sequoias caused Muir to advocate for further protection. He lobbied Congress for the act that created Yosemite National Park in 1890. The State of California, however, retained control of the park. Muir and his Sierra Club, along with other luminaries, such as landscape architect Frederick Law Olmsted, continued to lobby the government for an enlarged federal park. In 1903, Muir took President Theodore Roosevelt on a camping trip into the valley and persuaded him to put Yosemite Valley and the Mariposa Grove under federal protection. In 1906, Roosevelt signed a bill that placed Yosemite National Park under federal stewardship.

Muir was also an amateur scientist. He theorized that the major landforms in Yosemite were created by glaciers. Muir studied and described what he saw in terms of the recently proposed and, at the time, controversial theory of ice ages and glaciation. In 1871, he published his first article on glaciers in Yosemite in the *New York Tribune*. His style is exuberant:

> The great valley itself, together with all its domes and walls, was brought forth and fashioned by a grand combination of glaciers, acting in certain directions against granite of peculiar physical structure. All of the rocks and mountains and lakes and meadows of the whole upper Merced basin received their specific forms and carvings almost entirely from this same agency of ice. . . . Such was Yosemite glacier, and such is its basin, the magnificent work of its hands. There is sublimity in the life of a glacier. Water rivers work openly, and so the rains and the gentle dews, and the great sea also grasping all the world: and even the universal ocean of breath, though invisible, yet speaks aloud in a thousand voices, and proclaims its modes of working and its power: but glaciers work apart from men, exerting their tremendous energies in silence and darkness.[2]

These ideas were contrary to the conventional wisdom and theories of the geological establishment as championed by Josiah Whitney, head of the California Geological Survey. Whitney thought Yosemite Valley was not formed by glaciers, but that "the bottom of the Valley sank down" along faults. It turned out, in the fullness of time, that Muir was right and Whitney was wrong. Science progresses by fits and starts, encouraged by heated debate.

The park was designated a World Heritage Site in 1984 in recognition of its gleaming granite cliffs, foaming waterfalls, crystal-clear streams, giant sequoia groves, lakes, mountains, meadows, glaciers, and biological diversity. Almost 95 percent of the park is designated wilderness. The wilderness designation means that undeveloped federal land retains its wild character and is without permanent improvements or human habitation, so as to preserve its natural condition. One key aspect of its natural condition is the geology.

The oldest rocks in the Sierra Nevada were deposited along the west coast of **proto–North America** during the Precambrian and early Paleozoic. The west coast during the early Paleozoic was a passive plate margin; that is, there was no active deformation, much like the east coast of North America today. Starting in the middle Paleozoic (Late Devonian), perhaps 370 million years ago, a subduction zone evolved in the area of the Sierra and the Central Valley. This was the driving force behind the **Antler Orogeny**, the mountain-building event that uplifted and deformed the ancient rocks mostly in Nevada and Utah.* Sands and muds were dragged into the subduction zone and accreted, or pasted, onto the continent as a Franciscan-style metamorphic mélange, although this was roughly 150 million years before Franciscan subduction began. The remnants of this primeval accretion can be seen as metamorphic rocks on the eastern and western margins of the Sierra Nevada and in the roof pendants, bits of the original rocks preserved above and in the uppermost part of the granites.

Much later, around 230 million years ago, during the Triassic, a second subduction zone began off the west coast of North America. The magma generated by the down-plunging and melting oceanic crust and sediments

---

* *Orogeny* is the term geologists use for a mountain-building event, referring in particular to deformation and uplift as a result of folding and lateral compression of the Earth's crust. Some wags refer to these areas as orogenous zones.

rose buoyantly in the continental crust as city-sized blobs, or **plutons**, of molten rock; over a span of 150 million years, these plutons coalesced into a regionwide igneous mass, the Sierra Nevada **batholith**. Between 230 and 80 million years ago, the magmas moved up into the older rocks, enveloping some and melting and mixing with others. The early phase of these intrusions coincides with the **Nevadan Orogeny**; the later phase was part of the **Sevier Orogeny**. The timing is not terribly accurate, but the Nevadan Orogeny occurred along the western margin of North America mostly during Late Jurassic to Early Cretaceous time (approximately 160 to 145 million years ago). A continental **magmatic arc** (also called a volcanic arc), a string of volcanoes, developed due to subduction of the Farallon Plate beneath the North American Plate during the early stages of this mountain-building event. As subduction continued, several oceanic volcanic arc terranes are thought to have accreted onto the western margin of North America. The Sevier Orogeny, from approximately 145 million years ago to around 50 million years ago, was a continuation of this mountain building, but much farther east in Utah, Wyoming, Montana, and Alberta. It was characterized by eastward thrusting and folding, and it, too, was caused by subduction of the oceanic Farallon Plate beneath the North American Plate.

The granodiorite and granite magmas, cooling slowly at depths of around 6 miles, formed rock with large glassy-white crystals of quartz, feldspar (white, cream-colored, and pink), black **biotite** mica, and black **hornblende**. A belt of volcanoes erupted above the magma, spreading ash over much of western North America. Some of the peaks in this volcanic arc are thought to have reached up to 15,000 feet in elevation. This was what's known as the ancestral Sierra Nevada.

The volcanoes of this ancient Sierra died during Late Cretaceous time. One suggested reason is that the subducted

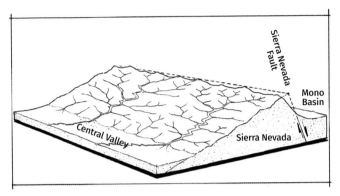

*Uplift and tilting of the Sierra Nevada block
(Adapted from Huber, 1987, p. 28)*

Farallon Plate flattened out under the continental crust, causing volcanism to migrate eastward. By early Cenozoic time, between 65 and 25 million years ago, the ancestral Sierra had been eroded to a low range of hills. As we have seen at Monticello Dam and Mount Diablo, sediments eroded off the ancestral Sierra were deposited in the ocean basin to the west, where they formed the Great Valley Sequence.

Sometime between 15 and 25 million years ago, when the Farallon Plate and mid-ocean spreading center had been completely subducted—that is, overridden by the North American Plate—a new interface developed between these massive crustal plates, a transform plate margin. Now the North American Plate was butting up against the Pacific Plate. But instead of simply converging, the Pacific Plate was moving northwest with respect to North America. The two plates slipped past each other along a transform fault, the San Andreas Fault system. Around the same time, stretching and rifting began in the Great Basin, that region from the Sierra Nevada east to the Wasatch Front in Utah. Rifting led to massive faulting along the eastern margin of the modern Sierra Nevada. The Sierra crustal block began to uplift and tilt west 5–10 million years ago. The uplift increased the stream gradient, causing intense down-cutting of rivers flowing west across the range and leaving some of the old stream channels abandoned on the upper ridges. The uplift continues today by way of multiple, mostly small-magnitude, episodic earthquakes. The average rate of uplift, as measured at Mount Dana along the **Sierra Crest** in Yosemite, is estimated at 1.5 inches per year. I experienced this gradual uplift myself while backpacking in the Sierra. Three of us were resting on a ridge south of Fremont Lake, just north of Yosemite Park, at around 3:45 p.m. on July 8, 2021, when we experienced a magnitude 6.0 earthquake. There was a loud noise like a distant explosion, followed by a sound like a low-flying aircraft approaching us from the north. The ground shook for about ten seconds, and as it did, we could hear the crack of rocks falling on both sides of the valley. Being on a ridge, we got lucky and had one bar of phone reception. We quickly logged in to the USGS *Did You Feel It?* website. As it happened, we were about 20 miles south of the epicenter.

Cascade Range volcanism, mostly in Washington and Oregon, reached as far south as Yosemite National Park. Ranging in age from 5 to 20 million years old, most of it is andesite lava (typical of the Andes, for which it was named). It is found mostly north of Tuolumne Meadows. Andesite, in contrast to granite, has more dark minerals: it derives from magmatic mixing of dark, dense oceanic basalt and light continental granite. An outcrop of 9-million-year-old **columnar basalt**, Little Devils Postpile, occurs along the Glen Aulin trail south of the Tuolumne River and a few miles west of Tuolumne Meadows. The most recent Cascade-related volcanism is a small, 3.5-million-year-old basalt flow just south of Merced Pass. These are small, localized outliers of the main **Cascade volcanic arc**.

The Spaniards, looking eastward across the Central Valley in the 1700s, saw *sierra nevada*, "sawtoothed snowy mountains." It is recent erosion that gives the range its rugged appearance. The snow is but a remnant of what had been.

Between 2 and 3 million years ago, the slowly uplifting granite block that is the Sierra had become high enough that, combined with global cooling, glaciers and ice sheets began to form. Glaciers form when more snow falls in winter than melts in summer. That's all it takes. Ice sheets are more or less stationary masses of ice, like what covers most of Antarctica and Greenland. Glaciers are the fingers of ice moving down from ice sheets. Moving glaciers sculpted and scoured the Sierra into the landscape seen today.

Although there have been dozens of advances and retreats of the glaciers over the past 3 million years, four major glacial episodes are recognized in the Sierra Nevada. From oldest to youngest, these are the Sherwin, Tahoe, Tenaya, and Tioga stages. The Sherwin was the largest, filling Yosemite Valley with 4,000 feet of ice and lasting 300,000 years. It was over by 1 million years ago. The later three stages are parts of the Wisconsinan glacial period that started around 130,000 years ago. The last major glacial episode, the Tioga phase, had its maximum extent 20,000 years ago and lasted to about 15,000 years ago. Since that time, glaciers have been largely retreating or absent in the Sierra. A few small glaciers formed around 3,000 years ago and lasted until the 1900s. The present glaciers, including McClure Glacier and Lyell Glacier, formed during the "Little Ice Age" (1200–1850 AD). At the present rate of melting, they are expected to be gone between 2030 and 2040.[3]

During the Sherwin stage, the longest glacier ran 60 miles down the Tuolumne River, even extending west of Hetch Hetchy dam on the western flank of the range. The thickest glaciers covered all but the highest peaks. Yosemite Valley was filled by the Merced Glacier.

Glaciers erode the landscape by plucking at rocks lying at their bases as well as at their sides, and scouring and abrading the rocks that they are moving over. The rocks thus removed are deposited as **moraines**, huge unsorted piles of rock dumped along the sides and front of the glacier as if a giant bulldozer had plowed the earth into furrows and ridges. We will see these ridges along the road between Tioga Pass and Lee Vining. Glaciers also create **till**, a loose mixture of glacial sand and powder derived from the grinding at the base of the ice. We will see glacial till in Tuolumne Meadows. Classic landforms carved by glaciers include U-shaped canyons (Yosemite Valley, Hetch Hetchy Valley); **hanging valleys** (above Yosemite Falls); large meadows that once were lakes (Yosemite Valley); glacially polished and striated granite; **arêtes** (sharp ridges); **horns** (glacially carved peaks); *rôche moutonnées* (asymmetric knobs with a gentle face in the up-glacier direction and a steep face on the downstream side; an example is Lembert Dome); **cirques** (bowls carved at the head of glaciers); and **tarns** (lakes within cirques).

The most recent moraines are on north-facing cirques; they contain fresh rock and little soil or vegetation cover, can be unstable, and still have ice mixed within the jumble of rock.

Tioga-stage glaciation, the last of four major glacial episodes (Adapted from Huber, 1987, p. 47)

The other landform that is characteristic of the High Sierra is the granite dome. These domes are caused by **exfoliation**, a type of erosion peculiar to granite.

When referring to people, exfoliation is the process of removing dead skin cells from the outer layer of skin. This is supposed to rejuvenate the skin and make it appear younger. When it comes to granite, exfoliation is the process of spalling off the uppermost layers of rock, like peeling a giant rock onion. When a mountain range is uplifted 5 miles and the rocky cover is eroded off, a lot of pressure is released from the remaining rock. The rock, having been unloaded, has a tendency to expand upward and outward, producing joints parallel to the surface. Because they are parallel to the surface, the jointed granite forms rounded domes. Half Dome, which as its name implies is a large granite dome (cut in half) at the east end of Yosemite Valley, is mostly massive granodiorite. It was created by exfoliation. The sheer face was gouged out by glaciers flowing down the valley.

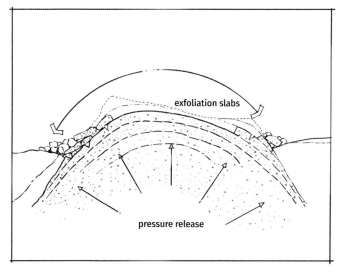

Progressive rounding of granite as the process of exfoliation spalls off successive sheets of granite (Adapted from Huber, 1987, p. 34)

YOSEMITE NATIONAL PARK AND YOSEMITE VALLEY

Yohhe'meti (Southern Miwok) or Yos s e'meti (Central Miwok) originally referred to the Indigenous group that lived in the valley. Yosemite literally means "grizzly bear" or "band of killers," and was used by the surrounding Miwok bands, who feared the group that lived in the valley. This name was adopted by the earliest Americans in the area.

Speaking of the California grizzly, there aren't any. Although a grizzly is on the state flag, and the golden bear is the mascot for UC Berkeley and the bruin for UC Los Angeles, the last hunted California grizzly bear was shot in Tulare County in 1922. In 1924, what was thought to be a grizzly was spotted in Sequoia National Park. After that, grizzlies were never seen again in California. So, less than 75 years after the discovery of gold, every grizzly in California had been killed. It is a sad tale, and worth telling.

Grizzly bears originally covered two-thirds of the state, from the northern forests to San Diego, pretty much all but the northeast and southeast corners. The bear was most abundant in chaparral, oak woodland, and hardwood forests. Many California place names reflect their former abundance: Grizzly Peak, Big Bear, Bear Valley, Bear Creek.[4]

At one time, the grizzly was plentiful. In 1602, Sebastián Vizcaíno sailed into Monterey Bay. He noted that the land teamed with ducks, geese, doves, quail, condors, elk, deer, and rabbits. He described grizzlies coming onto the beach at night to feed on a dead whale washed up on shore. This was the earliest clear settler account of a California grizzly. José de Galvéz led an expedition to explore California 160 years later. His men were crossing a swamp north of Point Conception when they saw tracks of a bear. They were hungry, so, armed with sword, lance, and musket, they tracked the bear down and killed it. Miguel Constanso, expedition cartographer, described the bear: "It was an enormous animal: it measured 14 palms from the soles of its feet to the top of its head; its feet were more than a foot long; and it must have weighed over 375 pounds. We ate of the flesh and found it savory and good."[5] This was likely the first California grizzly killed by Europeans. A few days later, they came upon a valley near the future site of San Luis Obispo and saw a large number of grizzlies digging for roots. These didn't die as easily, some taking up to nine bullets before falling. The Spaniards named the place La Cañada de los Osos, and it is still known as Valley of the Bears.

These were early days at the Monterey settlement, and the newcomers depended on ships bringing in supplies of food and goods from Mexico. The supply ships didn't always make it, and food was often scarce. Remembering the Valley of the Bears, the governor and thirteen soldiers returned there and spent three months killing bears and drying their meat. Father Palou wrote in his diary that they sent back "25 loads, or about 9,000 pounds of bear meat."[6]

In 1784, large land grants were given to Spanish officers, and they began running cattle on their ranches. The California grasslands became a grazing paradise. Cattle and horses were left to run wild by the thousands. Hides and tallow were harvested from cattle. During the butchering season, cattle were killed by the hundreds every day, stripped of their hides and rendered for tallow. The choice cuts of meat were saved for humans; the rest was left for the vultures, wolves, and grizzlies. Horses, too, multiplied rapidly, but because they competed with cattle for grass, they had to be culled. Their carcasses were left to rot, adding to the grizzly's food sources. Because cows were slower than elk and deer, it didn't take long for grizzlies to learn to kill cattle. Consequently, there was a population explosion among grizzlies: it is estimated that there were ten thousand in California in the early 1800s.[7] John Bidwell spotted a group of sixteen grizzlies in the Sacramento Valley in 1841, stating that "grizzly bears were almost an hourly sight, in the vicinity of the streams, and it was not uncommon to see thirty to forty a day."[8]

To eliminate the bears, Spanish ranchers used a poisoned "bait ball" made of suet or pig entrails filled with strychnine, which they hung from the branches of a tree. Mexican settlers captured bears for bear and bull fights. Roping the big bears became a favorite sport of the Spanish, and later of the Mexican cowboys. They would lasso a bear and then kill it with a machete, sword, or lance. An American by the name of Davis married a Spanish California girl and told how his father-in-law, with the help of ten soldiers and a string of horses, "lassoed and killed 40 bears in one night." The bears had come down from the mountains to partake of the remains of cattle that had been slaughtered for their hides.[9] Don José Ramón Carillo, an aristocrat who hunted grizzlies for sport, was famous for his single-handed duels during which he faced and dispatched bears with a light rapier, many in the San Fernando Valley outside of Los Angeles.[10]

The grizzly became a symbol of the Bear Flag Republic, the short-lived attempt by a group of American settlers to break away from Mexico in 1846. This rebel flag became the basis for the state flag of California, after which California became known as the Bear Republic.

In 1866, a grizzly bear described as weighing as much as 2,200 pounds was killed in Valley Center, 40 miles north of San Diego. The incident was recalled in 1932 by Catherine Lovett Smith, who witnessed the killing on her family's ranch when she was six years old. If the measurements were accurate, this particular bear was the biggest ever found in California.[11] Other sources confirm Lovett Smith's account of the bear, but differ as to its size.

The cattle population was greatly reduced after extreme drought and flooding in the 1860s and 1870s, and this led to a decrease in the number of grizzlies. This coincided with an increased interest in bear hunting, both for sport and commercial purposes. Grizzly bear meat became common on restaurant menus in the San Gabriel area; according to Mike Davis, "The paws from adult bears and the flesh from young cubs were deemed particular delicacies."[12] Until the bear went extinct, American settlers paid bounties on the bears that preyed on livestock. Absolom Beasley, for example, hunted grizzly bears throughout the Santa Lucia Mountains and claimed to have killed 139 bears in his lifetime. The California mountain man Seth Kinman claims to have shot more than eight hundred grizzly bears in a 20-year period in the area around Humboldt County.[13] So it is no wonder that the magnificent animal is gone.

I have hiked among grizzlies in the Canadian Rockies and can attest that there is something both alarming and magical about coming on a fresh 12-inch paw print on the trail. All your senses are instantly on high alert. Perhaps one day the grizzly will again be seen in the California wilderness, much as the mountain lion, tule elk, and bighorn sheep are making a comeback. I, for one, look forward to it.

Getting back to geology, the massive granite of the Sierra gives Yosemite Valley its towering 3,500–4,000-foot-high walls, scenic domes, and clear streams. This tough rock, when eroded, weathers to a clean sand rather than mud, so the water running off the cliffs remains crystal clear. The granite seen in Yosemite Valley is the lower part of the volcanic arc that formed the ancestral Sierra Nevada between 210 and 80 million years ago. The primary rock in Yosemite Valley, as in most of the Sierra Nevada, is granodiorite. There are also varying amounts of diorite (a darker rock with little quartz), **tonalite** (containing more feldspar), and real granite (containing more pink feldspar).

The oldest rock in the Yosemite Valley area is exposed at the west end of the valley and along Route 140, El Portal Road. It includes diorite, tonalite, and granodiorite. The granodiorite can be seen east of the Arch Rock Entrance Station. The darker, highly jointed diorite forms the Rockslides adjacent to El Capitan and the dark **dikes** (narrow vertical intrusions) on the east face of El Capitan. Light-colored tonalite can be seen at the Gateway along El Portal Road in the Merced Gorge east of El Portal. The tonalite has been dated at 114 million years old, and all three of these rock types are assumed to be around that age.

The 108-million-year-old El Capitan Granite was injected into these older rocks and is the dominant rock in the west half of the valley. As the name suggests, the sheer cliffs of El Capitan are made of massive El Capitan Granite. The Half Dome Granodiorite, at 87 million years old, is the youngest **intrusive rock** in the valley and forms the cliffs around Mirror Lake and at Half Dome.

But beyond the rocks themselves, Yosemite is the poster child for glaciation. The valley has that classic U-shape, with a flat bottom and steep walls, caused by a glacier grinding into the valley sides as it moved downslope. Once the main glacier melted, tributary streams that flowed into the main valley were left stranded at the top of the main canyon walls. The tributary valleys end suddenly at a steep cliff where they meet the larger valley. These are called hanging valleys, with some of the most beautiful waterfalls occurring where they enter the main valley.

I stop in Yosemite Valley only briefly, because I've been here many times before. If this is your first time, I encourage you to explore the valley at your leisure, for it is surely one of the wonders of the world. Be on notice that, because there are so many who wish to do the same, a system of permits has been instituted. See the National Park Service website to learn more: https://www.nps.gov/yose/plan yourvisit/permitsandreservations.htm.

To get from El Dorado County Historical Museum to Tunnel View, Yosemite, return south on Placerville Drive and merge onto US-50 West to Sacramento; take Exit 44A to Missouri Flat Road and turn left (south); turn right (south) onto Forni Road; turn right (west) onto Pleasant Valley Road, then immediately left (south) onto CA-49; as you approach San Andreas, turn right (south) onto Pool Station Road; turn right (south)

Sierra Nevada 127

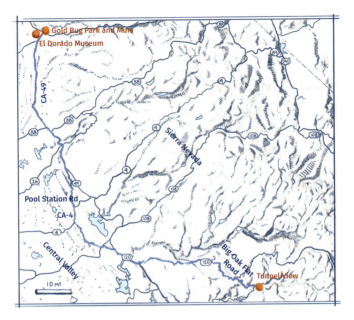

Map of the route from Gold Bug Park and Mine to Yosemite Valley

onto CA-4 and drive to Copperopolis; turn left (south) onto Main Street/E-15/O'Byrnes Ferry Road; turn left (east) onto CA-120/ Big Oak Flat Road; once in Yosemite, bear left on El Portal Road, which becomes Southside Drive, then turn right onto CA-41/Wawona Road and drive to the Tunnel View pull-out on the right (37.715680, −119.677093), for a total of 141 miles (3 hrs 25 min).

STOP 21: TUNNEL VIEW

I'm driving along a winding mountain road, CA-120, and come to the ranger station at the entrance to Yosemite. There is the seemingly ever-present road construction and some hard-core cyclists, so traffic moves slowly. It smells like, um, nature, as the sun heats the pitch in the tall pines. There are probably forty cars ahead of me, and it takes a good twenty minutes to get to the ranger booth, where I pull out my

Yosemite Valley II (*Woodcut © Tom Killion. Reproduced with permission.*)

Lifetime Senior National Parks Pass.** We get through the traffic logjam and drive another few minutes. So far nothing out of the ordinary, just driving through thick forest with tall pines and fir trees, and manzanita chaparral in the old fire scars. I drive past the Crane Flat campground, where the Tioga Pass Road heads east across the mountains. I've camped here before, and I can affirm that it is a nice spot to spend a night, especially in the offseason when it's not as crowded. Beyond Crane Flat there is a good parking area at Yosemite Valley Vista, but I can just barely get a glimpse of El Capitan and Half Dome behind a ridge. No, Tunnel View is definitely the place to go.

The road descends to 4,000 feet at the floor of Yosemite Valley. The valley floor, here at the west end, contains mixed pine and hardwood forest. It's dark beneath the overlapping canopies of black oak, ponderosa pine, incense-cedar, and white fir. The lane crosses the Merced River, a foaming torrent in springtime and gently flowing the rest of the year,

---

** By the way, this is the best deal around, hands down. If you are old enough to qualify (62 or older), get one. There has to be some advantage to getting older.

Valley of the Yosemite, *an oil painting in the style of the Hudson River School, by Albert Bierstadt. This 1864 painting is an idealized and romantic representation of a frontier landscape, and one of my favorite evocations of nature.*

meandering through the trees. I make a hard right turn onto Wawona Road and head back toward Fresno, but only go 1½ miles and pull over on the right at the Tunnel View turnout. My first time here I was facing away from the valley as the road climbed to this spot, so I wasn't paying attention to the view. There are lots of cars and tourists here. I got out of the car and stretched, wondering what all the fuss was about. I turned to take in the view and there it was, laid out in all its glory: Yosemite Valley.

Every time I stop here, I'm gobsmacked all over again. Yosemite Valley is the Sierra through and through: glacier-polished granite, stunted trees on the ridges, deep-blue sky. Framed by the imposing rock face of El Capitan on the left and delicate Bridalveil Fall on the right, and with iconic Half Dome as a backdrop, this vista has been painted and photographed for generations. Yosemite Valley, flat bottomed, green, and serene, is embraced by towering cliffs on all sides. Giant trees form dark groves, and the crystalline

Merced River wanders across sun-dappled meadows carpeted with wildflowers.

Thunderheads rise above it all. It is Earth's Disneyland, a place of inspiration, a place for reflection, a place to marvel at the grandeur of nature. And it comes with a two-mile queue of cars and an annual crowd of tourists that exceeds the population of Canada. I exaggerate only a little: 3.9 million tourists visited the park in 2023.

From this vantage point, I have often descended into the heart of the valley. There, trails lead to many of the waterfalls, or wander along the banks of the Merced River or to the base of El Capitan and other world-renowned destinations for rock climbers. One of the first things I notice when I get into the valley is that I can't just cast my eyes upward to see the valley rim. It is so steep that I have to physically tilt my head back. On the massive rock faces there is no sense of scale. Bystanders set up telescopes to watch ant-like climbers perched precariously on the face of El Capitan.

Yosemite Valley landmarks (From Huber, 1987, p. 18)

To get from Tunnel View to Olmsted Point, backtrack down Wawona Road into Yosemite Valley and bear slightly right onto Southside Drive; turn left onto CA-41/CA-120/El Capitan Drive, and continue straight onto Northside Drive (there are some great views of El Capitan and Bridalveil Fall along this stretch); turn right onto Big Oak Flat Road; at Crane Flat turn right onto CA-120/Tioga Road and drive to Olmsted Point on the right (37.810857, −119.484556), for a total of 45.2 miles (1 hr 12 min).

THE HIGH COUNTRY

On this occasion I leave the valley and continue into the high country, a place of alpine lakes, rugged peaks, and magnificent meadows. This is where many backpackers begin the John Muir Trail, a highland adventure that stretches 220 miles south to the summit of Mt. Whitney. It's a place where, if I've timed it right, I can leave the crowds behind and find solitude amid the mountains.

STOP 22: OLMSTED POINT AND TENAYA LAKE

This viewpoint is named for Fredrick Law Olmsted, designer of New York's Central Park and landscaper of the 1893 Chicago World's Fair; the University of California, Berkeley; and Stanford University, among many other places. Not that this view needed designing: nature's knobs and domes are not so much architectural and designed as they are naturally curvaceous and even sensual. Still, it's a nice tribute to the man.

At 8,420 feet above sea level, the High Sierra is all around me now. I am near timberline, and the trees are thinning out. This is the first time I can look in any direction and see the smooth and rounded domes of exfoliating granite that are characteristic of the high country. The rock at this pullout is Half Dome Granodiorite, and it has excellent exfoliated sheaves of granite and examples of **glacial polish**. Yes, ice can polish rocks, even rock as hard as granite. Viewed at the

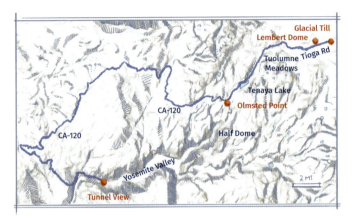

Map of the route from Yosemite Valley to Olmstead Point and Tuolumne Meadows stops

correct angle in relation to the sun, the granite has a mirror-like sheen. The polishing was done by glaciers moving over the rock and grinding it, just like polishing a mirror, with extremely fine grit entrained within the ice.

The valley below this spot contains Tenaya Creek, flowing south to Yosemite Valley. As I look south down the valley, the back side of Half Dome is plain to see.

I grab my camera and walk to the far east end of the parking area. I gaze northeast toward the Sierra Crest. From here it is just possible to make out the sky-blue gem of Tenaya Lake surrounded by multiple granite domes and, if the weather is right, towering cumulus clouds. Tenaya Lake was gouged out of Half Dome Granodiorite by the Tenaya Glacier on its way from the high Tuolumne ice field down to Yosemite Valley. Poly Dome is the prominent exfoliation dome to the left (north) of Tioga Road. Dark vertical stripes mark the various domes' flanks, water stains from snowmelt trickling off their summits. The smooth, sculpted Rubenesque granite voluptuously bends and curves like the graceful lines of a giant fertility figure. This is nature's artistry, shaped over eons, a breathtaking landscape that lifts the spirit and inspires the soul.

The view from Olmsted Point looking northeast. Poly Dome (left), Mount Conness, Pywiack Dome, and Medlicott Dome (right). Tenaya Lake is in the foreground.

When the Ahwahnechee led by Chief Tenaya fought the Mariposa Battalion, the American soldiers met with Tenaya near the lake and named it after him. But the original, Indigenous name is Pywiack ("Pie-we-ack"), meaning "Lake of the Shining Rocks." And it is.

This, in my opinion, is one of the most scenic spots in the Sierra. But don't just take it from me. The lake and Poly Dome form the backdrop for one of Ansel Adams's most beloved and well-known photographs, his 1946 black-and-white *Tenaya Lake*.

To get to Tuolumne Meadows from Olmsted Point, continue east on CA-120/Tioga Road for 10.3 miles (20 min) to the Lembert Dome picnic area on the left (37.877269, −119.353244).

STOP 23: LEMBERT DOME AND TUOLUMNE MEADOWS

Tuolumne Meadows reveals itself in a rush. I am heading east on the Tioga Road surrounded by forest. Then the expanse of grassland and the lazy Tuolumne River open before me, the turf seeming to glow in the high-altitude light. It is early morning, and I'm lucky: herds of mule deer, renowned for their outsize ears, are browsing across the grassy field. The rocky wedge of Lembert Dome rises out of the meadow to the east.

My first recollection of Tuolumne Meadows is as a sixteen-year-old completing the last stretch of the John Muir Trail. My two companions and I had just topped Donohue Pass, and we looked down on the headwaters of the Lyell Fork of the Tuolumne River. We were young and strong, but still the meadows seemed to go on forever. The valley is framed by peaks that poke above timberline, peaks and ridges with sonorous names like Vogelsang Peak and the Kuna Crest, many with glacial cirques and ice fields on their north slopes. Anticipating the end of our trek, like horses sensing the barn, we plunged forward.

Together with the meadows along the Lyell Fork, Tuolumne Meadows, at an elevation of 8,600 feet, is the essence of an alpine pasture, up to 1 mile wide and stretching

Lembert Dome at Tuolumne Meadows. The Tuolumne Ice Sheet moved from right to left (toward the north).

over 10 miles. It contains lazy meandering streams with crystal-clear swimming holes just begging you to go for a dip. And some of the bluest skies anywhere. Except in the afternoon, when they frequently play host to building thunderheads.

There are two types of material underlying Tuolumne Meadows: granodiorite and glacial till. The granodiorite forms the Tuolumne Intrusive Suite, consisting of four intrusions stacked one inside the other, like those nested Russian matryoshka dolls. Over millions of years, progressively younger magmas were inserted into older granites. They are all Cretaceous in age. From oldest to youngest and outer to inner, they are Kuna Crest Granodiorite (about 91 million years old), Half Dome Granodiorite (87 million years), Cathedral Peak Granodiorite (86–88 million years),

Tuolumne Meadows from Pothole Dome
(Woodcut © Tom Killion. Reproduced with permission.)

and Johnson Granite Porphyry (85.4 million years). Most of the differences between these rocks would only be noticed by a geologist specializing in granites. This would be like noticing the difference between chocolate chip cookies with and without oatmeal. Lembert Dome, before us at this stop, is made of Cathedral Peak Granodiorite. Look for the large orange-pink feldspar crystals.

Lembert Dome is a favorite of both hikers and climbers. Climbers, obviously, like the steep side. I like walking up the gentle back side of the dome, gaining about 800 feet of elevation, to get a view over the meadows. The cool thing about this dome is its shape: it is a classic roche moutonnée. This is a landform created when glaciers move over a preexisting hill. The glacier grinds the hill into an aerodynamic shape where the up-glacier direction is the gentle slope, and the downstream side is the steep slope. Oddly, roche moutonnée is French for "rock wigs." In 1787, the alpine explorer and geologist Horace Bénédict de Saussure thought these whale-back ridges resembled the wavy wigs or *moutonnées* that men wore at that time, wigs that were slicked down with mutton tallow to keep them in place. The dome was named after homesteader Jean Baptiste Lembert, who lived here in the mid-1860s.

Sandwiched between the grass of the meadows and the granite lies a veneer of debris, the other type of rock in the Tuolumne area, glacial till. We meet the till up close at the next stop.

To reach the Glacial Till stop from Lembert Dome, drive 1.5 miles (3 min) east on CA-120/Tioga Road and pull over on the right shoulder (37.880207, −119.328003). The roadcut through glacial deposits is on the north side of the road.

STOP 24: GLACIAL TILL

A short stretch up the road from Lembert's giant granite dome is the other stuff that is characteristic of the uplands: glacial till. There is a nice roadcut that exposes an approximately

Sierra Nevada 137

Glacial till in roadcut, Highway 120 (Tioga Road) just east of Lembert Dome

30-foot-high section of this glacial material. In the roadcut it is obvious that a large range of sizes and shapes of rocks make up the glacial sediment. Glacial till typically consists of white powdery rock flour, silt, sand, cobbles, and boulders. It is the stuff that was plucked and scoured out along the base of moving ice and left behind as the glacier retreated. Tuolumne Meadows is filled with glacial till deposited by the Tuolumne Ice Sheet when it melted some 12,000–15,000 years ago. The ice sheet is thought to have been up to 2,000 feet thick in this area. Imagine the immense weight and power it must have had, grinding downhill under the influence of gravity, as it slowly and inexorably sculpted the landscape.

Glacial till, this humble, loose gray dirt that was ground out of peaks and ridges, bears witness to the enormous and dynamic forces that fashioned the Sierra high country.

To get to Mount Dana View, continue driving east on CA-120/ Tioga Road 5.3 miles (10 min) to a parking area on the right (37.908120, −119.258502) just inside the Tioga Pass park entrance.

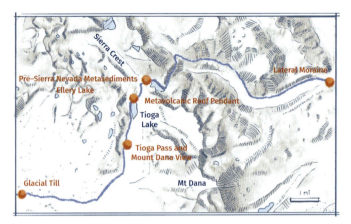

Map of the route from Tuolumne Meadows to Lee Vining

STOP 25: TIOGA PASS AND MOUNT DANA VIEW

As I continue east on Tioga Road, the forest closes in around me again. But it is a thinning, high-altitude forest of lodgepole and yellow pines. The climb is gentle on this side of the pass: I gain about 1,000 feet in the five miles between my last stop and the pass. And because timberline is around 9,600 feet here, as I approach the pass I break out of the forest, and the highlands are laid out before me. I am climbing the Sierra Crest (also known as the Sierra Divide), that high ridge and **drainage divide** separating streams that flow west into the Pacific from streams that flow east into the Great Basin. Snow can be found in shady spots along these ridges almost year-round.

Tioga Pass is a **col**, the lowest point on a ridge between two peaks. This particular col, at 9,943 feet, happens to be on the Sierra Divide, and was formed by glaciers moving away on either side of the ridge, cutting and carrying the rocks away with them. Tioga Lake and the other large ponds nearby are **kettles**, formed as the glacier was melting and retreating. Blocks of ice left behind became isolated in the glacial till and melted, leaving shallow depressions that filled with water.

North of here, the Sierra Crest is lower and has passes that are suitable for roads in several spots. Donner Pass and Carson Pass, among others, were blazed by the forty-niners and settlers coming to California. Interstate 80 climbs to Donner Summit in wide, sweeping curves that are two and three lanes wide in each direction. By contrast, Tioga Road even today tightly hugs the contours of the hills, one lane in each direction, and has guardrails only in the steepest sections. South of Tioga Pass, on the other hand, is the High Sierra. The mountains are so tall and rugged that no highway crosses the range for another 150 miles. The High Sierra includes Mount Whitney, at 14,505 feet the highest peak in the lower forty-eight states. Tioga Road (California Route 120) is closed in winter due to the altitude of the pass and abundant snowfall.

Tioga is an Iroquois word meaning "where it forks," but there were no Iroquois around here; the Iroquois Nation was largely focused around the Great Lakes. However, there were miners who came from Back East. Tioga Road is the legacy of a failed nineteenth-century mining effort. In 1880, two prospectors working for the Swift Brothers of Chicago discovered a silver-bearing vein in the area and named it Tioga. They excavated a 4,000-foot-long tunnel along the vein. A road was needed to get equipment in and ore out. Tioga Road up the eastern side of the Sierra was started in 1882; it took five months to build and cost $62,500, a fortune at the time. (This would be about $2 million in 2023 dollars.) After investing a total of $312,500 to develop the mine, Swift Brothers suddenly ordered that all work be stopped; equipment was pulled out, and the mine was abandoned. The vein was never produced. In 1915, the Department of Interior bought the road and opened it to the public. It is famous for being very winding and steep and for offering breathtaking vistas, as I was about to experience again as I continued driving east.

Mount Dana as seen from the Tioga Pass park entrance

I stop at the pullout on the east side of the highway just south of Tioga Pass and hard by the entrance station. The featured view is south toward Mount Dana on the Sierra Crest, a landmark in these parts. The mountain and rocks east of this stop are reddish-brown Mesozoic metavolcanics that were part of the pre–Sierra Nevada (Triassic, roughly 222-million-year-old) continental-margin volcanic arc. These volcanics were later folded and pushed eastward by thrust faulting during the Jurassic, and were later metamorphosed when the Sierran granitic magmas invaded them during the Cretaceous.

The 13,061-foot peak is named after James Dwight Dana, who taught geology and natural history at Yale from 1850 to 1892. It was named by California State Geologist Josiah Whitney when he climbed the peak in 1863. Dana was so well thought of that it is worth digressing for a moment to describe the man. Dana did pioneering studies of mountain-building, volcanic activity, and the origin and structure of continents and oceans. He was a mineralogist and zoologist as well as a geologist, and traveled widely, including to Hawaii, the Andes, and Mount Shasta in California. He devised a system of organizing minerals based on their chemistry and **crystal system**, or crystal shapes. *Dana's System of Mineralogy* was published in 1837, and geology students still use his system to classify minerals. He became the editor of the *American Journal of Science and Arts* and was named president of the Geological Society of America in 1890. The Royal Society (of the UK) awarded him its Copley Medal in 1877; the Geological Society of London gave him a Wollaston Medal in 1874; and the Royal Society of New South Wales, a Clarke Medal in 1882. Clearly, he was the preeminent American geologist of his time. Ironically, he never made it to Yosemite.

I love the mountains with their robin's-egg-blue sky, crystal-clear water, and cool dry air. As I stand at the eastern entrance of Yosemite Park, Mount Dana to the south, the view ahead into the Great Basin is daunting. What lies before me shimmers in a heat haze, a land of brown desert valleys and gray-green ranges extending one after another to the horizon. The Earth is literally pulling itself apart, leaving in its wake long, north–south-faulted mountains separated by long, north–south fault-bounded basins. From Tioga Pass, at the western margin of the Basin-and-Range, the view is punctuated by the Sierra Nevada Fault that forms the massive escarpment of the eastern Sierras. I hesitate a moment, trying to take it all in, before continuing onward.

CHAPTER 5

Basin-and-Range

The Basin-and-Range geologic province is massive. It extends from eastern Oregon and California, through southern Idaho; covers all of Nevada and western Utah; and extends to southern Arizona, northern Mexico, and southern New Mexico. Throughout all of this vast area, the Earth is extending. The extension has left the ranges—originally a high plateau—several thousand feet above the valleys, which have dropped down and filled with sediments. The valleys tend to be broad and flat, and often contain dry lakes, called **playas** from the Spanish word for "beach."

Within the Basin-and-Range, and as a result of it, lies the Great Basin. The Great Basin is distinguished by rivers that drain internally. There is no outlet to the sea.

I've backpacked in some of the ranges and can attest that the water is clear and fresh while in the mountains. Like most mountain streams, the water comes from the previous season's snowmelt. All of the major rivers end in low spots such as Great Salt Lake, Carson Sink, Humboldt Sink, Walker Lake, and Pyramid Lake. As the streams enter the arid valleys, there is no more water to add, and they become condensed by evaporation and by the irrigation ditches that siphon water to local hayfields. What is left is full of minerals, a brine that puddles in the valley bottom during spring runoff and then dries up to form a salty crust on the valley floor. It is always a surprise to me to see the abundance and diversity of birds that gather in these grassy oases, these briny remnants of once-vigorous mountain streams. The birds don't seem to mind; in fact, they thrive on the bugs that swarm in the salty mud.

The mountains rise abruptly along faults that bound the ranges. Ranges bounded only on one side by faults are tilted away from the high, faulted side, just like the Sierra Nevada. My personal preference when hiking is to climb the gently sloping, nonfaulted side. When I explored for minerals in these ranges early in my career as a geologist, I didn't care which side we went up, because we used helicopters.

Whereas the desert ranges tend to be a cool escape from the heat, the valleys are generally hot, except in winter, when they get cold enough to freeze a polar bear. Both livestock and wildlife gather at those places where springs issue from the broken rocks along the valley-bounding faults, places where groundwater emerges at the surface. Lots of water wells and a few oil wells have been drilled within these valleys. As the wells penetrate the valley fill, the drillers examine the rock chips brought to the surface. From these cuttings we know that the same rocks exposed in the ranges lie deeply buried beneath the valleys, below thousands of feet of **alluvium** that has been eroded from the adjacent mountains.

To a geologist, a pattern usually indicates a process. The pattern of north–south ranges and valleys indicates that the Earth's crust across the entire region is extending in an east–west fashion. We can tell when the faults began to move by dating the age of the oldest valley-filling sediments. From this we know that extension began in the northern Basin-and-Range during the Eocene, roughly 48 to 41 million years ago. The main phase of faulting is younger to the south, meaning that the extension began in the north and migrated southward. In Southern California, Arizona, New Mexico, and northern Mexico, faulting began from about 26 to about 15 million years ago. During late Tertiary time, extension appears to have migrated outward, west toward the Sierra and east toward the Colorado Plateau. The Colorado Plateau is the unusually thick and geologically stable section of the Earth's crust that extends from northern Arizona across eastern Utah into western Colorado and parts of northwestern New Mexico. It includes all of the flat-lying or gently tilted rock layers seen in the Grand Canyon, Zion Canyon, Bryce Canyon, Canyonlands, and Arches National Parks. Whereas the Colorado Plateau sits on crust that averages 30 miles thick, and average continental crust is around 25 miles thick, the extended crust in the Basin-and-Range province is 17–21 miles thick. When the Earth's crust extends, it gets thinner, like stretched taffy. Thin crust lets more heat escape from the mantle, which is why there are lots of hot springs and even volcanic rocks in places. Overall, the crust in the Basin-and-Range province has extended about 100 percent, meaning that the distance from Reno to Salt Lake City is twice what it would have been 50 million years ago.

There is a well-worn saying in the geosciences: if there are two geologists, there are at least three opinions. The scientific method famously incorporates the method of multiple working hypotheses. This means that we try to look at all the possible reasons for our observations and then eliminate them one by one until we come to the one explanation that best fits all the data. This method seems to follow Sherlock Holmes's admonition to Doctor Watson in *The Sign of the Four*: "How often have I said to you that when you have eliminated the impossible, whatever remains, *however improbable*, must be the truth?" And so there are many theories for why the Basin-and-Range province is extending, and geologists are actively working to see which is the best fit for what we see.

One possible explanation is that late Cenozoic regional uplift, extensional faulting, and basaltic volcanism are a result of the underlying mantle pushing up the crust and heating it. According to this hypothesis, convection in the Earth's mantle causes the overlying continental crust to bulge upward, with faulting, high heat flow, and volcanism as a result.

Another working hypothesis considers that compression and shortening during multiple mountain-building events (the Antler, Sonoma, Sevier, and Laramide Orogenies), along with injections of granite into the crust, led to overthickening and uplift of the crust. The bulked-up crust would have been over 36 miles thick, making Nevada and Utah a high plateau. When crust is that thick over a long period of time, it eventually becomes unstable, leading to gravity-induced collapse and extension outward from the center.

Yet another hypothesis given to explain the regional extension seen in the Basin-and-Range province involves changes in motion between the Pacific and North American tectonic plates. For the past 30 million years, the Pacific Plate has been moving slowly northwest relative to the southwest-moving North American Plate. The most obvious result of this is the San Andreas Fault, where the two plates grind sideways past each other. But the Pacific Plate has cast its influence across most of California, and the plate boundary is diffuse rather than a single break. In this scenario, the Central Valley and Sierra Nevada are being tugged gently to the northwest. As they move northwest, the Basin-and-Range province is slowly being pulled apart.

Another proposed explanation for the observed extension is **slab rollback** beneath over-thickened crust. According to this hypothesis, when the Farallon Plate was subducted beneath the North American Plate, for some reason not fully understood the subducted plate remained nearly horizontal, moving east beneath the crust rather than plunging back into the mantle, which would be the normal case. This **flat-slab subduction** caused regional compression that moved from west to east during the Sevier and Laramide Orogenies in Utah, Colorado, Wyoming, and Montana. The double thickness of crust created a high plateau extending from California to Colorado. The Sevier Orogeny was characterized by eastward thrusting of old continental-margin sedimentary layers from Utah to Alberta starting around 145 million years ago. The **Laramide Orogeny**, farther east and younger than the Sevier mountain-building event, occurred between roughly 80 and 55 million years ago. Laramide deformation is characterized by **basement-cored vertical uplifts** in Colorado, Wyoming, and Montana. These uplifts involve the basement rock of the continent being squeezed upward like a watermelon seed between your fingers. Eventually, when the deformation reached as far east as New Mexico and Montana, the subducted oceanic plate began to sink into the mantle, first in the east and then progressively westward (the slab rollback), pulling the overlying crust apart in the process.

While there is no hard-and-fast agreement as to why this region is extending, satellite-based GPS measurements confirm that extension continues at rates between 4 and 47 inches every 100 years.

THE EASTERN SLOPE

As I leave Tioga Pass and begin making my way down the eastern slope, it becomes abundantly clear that the Sierra Nevada is a tilted mountain range with a gentle western slope and a steep eastern side. The precipitous drop on the east is a result of a zone of roughly north–south-trending faults that raised and tilted the mountain range at the same time as they lowered the basin to the east. The steep eastern side is referred to as the **Sierra escarpment**. The scenery is grand, but the driving is hazardous due to the sheer drops on the side of the road. There are guardrails—in some places. Regardless, my wife is constantly reminding me to "watch the road, not the rocks." The road runs down Lee Vining Canyon, a valley carved out by glaciers.

I do recall the first time I was on this road. I was a terrified ten-year-old passenger in my parents' car on a family vacation to Yosemite and Lake Tahoe. I was most assuredly *not* looking at the geology. What I remember most is the precipitous drops at the road's edge. My dad explained that he had to use the engine to slow the car in order to keep the brakes from overheating. I don't think there were guardrails and pullouts in 1962, and the road had a notorious reputation as a killer. The trip down from the pass seemed much longer than it actually is.

Other than being steep and narrow, and only about six miles wide along CA-120, the eastern slope is much drier than the western slope because it lies in a rain shadow. The gentle rise along the western slope of the Sierras forces most of the moisture out of the air, and it falls as rain or snow before getting to the east side. The next three stops are in the transition zone between the Sierra Nevada and Basin-and-Range. Geologically, the escarpment is composed of the same materials as the bulk of the Sierra, Cretaceous granodiorites that invaded older metavolcanics and metasediments. The metavolcanics are exposed in a roadcut at the next stop.

To get from Mount Dana View to the roof pendant, drive 1.5 miles (3 min) east on CA-120 to the roof pendant at the Tioga Lake parking area on the right (37.928958, −119.254456).

STOP 26: SIERRA NEVADA METAVOLCANIC ROOF PENDANT AT TIOGA LAKE

From the pullout at the north end of Tioga Lake, I cross the road and walk about 1,500 feet northeast to the start of the Nunatak Nature Trail on the north side of the road. There is also a parking area here for westbound traffic. At first glance, this outcrop is just a rusty-brown and chalky-white jumble of rock. It takes a moment before I can even make out the layers inclined steeply to the southwest. The colors are the result of iron staining and leaching during metamorphism and later weathering.

Part of a roof pendant consisting of Mesozoic metavolcanic rocks is exposed in the roadcut. Recall that a roof pendant is a remnant of the rocks that were above a magma chamber, rocks that were heated and metamorphosed by the upward-moving granite. These early Mesozoic volcanic rocks are called the Saddlebag Lake pendant, after nearby Saddlebag Lake, and they are quartz-rich rhyolite volcanics that are partly older than and partly the same age as the granite found below them. The widespread volcanic ash-flow deposits near Saddlebag Lake have been dated at about 219 to about 232 million years old (Triassic). Thus volcanism began earlier than, and continued during emplacement of, the light-colored granite exposed east of here in Lee Vining Canyon. This granite, although separated by millions of years in time, has the same composition as the younger, Late Cretaceous Tuolumne Intrusive Suite that I saw in Tuolumne Meadows.

The age of this volcanism and the deeper magma it came from tells me that subduction was active along the western edge of North America during the Triassic. These are some of the earliest rocks formed as part of the volcanic arc related to subduction that was taking place all along the continent–ocean boundary west of here. The US Geological Survey has mapped these rocks and determined that, after eruption, they were intensely folded and metamorphosed during the Late Cretaceous, which would account for the steep inclination of the layers.

It's getting late in the day, and my stomach is rumbling. Back in Tuolumne Meadows, the grill closed at 5:00, and the Tuolumne Lodge wouldn't seat me until 8:00, so I continue east to the stops around Tioga Pass. Two miles outside the east entrance to the park, between Tioga Lake and Ellery Lake, I find the Tioga Pass Resort. There are a handful of cabins and a small, old-time diner. I can't vouch for the cabins, but the place has the best meatloaf I've eaten in years, and I finish it off with their blueberry-peach cobbler and ice cream. Hard to beat that.

Driving another mile east, watching the rocks rather than the road, I notice that I've left the metavolcanics and entered metasediments.

Metavolcanic roof pendant in a roadcut at Tioga Lake. I can just barely make out the layering inclined about 50–60 degrees to the left (southwest).

To get from the roof pendant at Tioga Lake to the metasediments at Ellery Lake, drive another 1.1 miles (2 min) east on CA-120 to the metasediments roadcut at the turnout to Ellery Lake on the right (37.938639, −119.244452). Be careful crossing the highway.

STOP 27: PRE–SIERRA NEVADA METASEDIMENTS, ELLERY LAKE

I pull over at the Ellery Lake campground turnout. Another outcrop of roof pendant above the Sierra granite is exposed here. This one consists of much older Paleozoic sedimentary rocks that existed long before the granitic rocks intruded and changed (recrystallized) them. The dark rock is a **hornfels**, a metamorphosed mudstone, and it is interlayered with **quartzite** (metasandstone) and marble (metalimestone). Hornfels is generally a dark and featureless metamorphic rock that has been altered by heat from nearby magma. This alteration, however, did not eliminate the original layering of the rocks.

It is next to impossible to date these metasediments because they lack recognizable fossils. They are considered to be Middle Cambrian to Middle Permian in age, a 270-million-year span of time that encompasses almost the entire Paleozoic Era. The eastern Sierra Nevada roof pendants are thought to be separated from all other rocks of the same age by faulting. The faulting formed a discrete structural block that has been referred to by some as the Morrison block. Within the Morrison block there are multiple roof pendants, including the Saddlebag Lake pendant in this area and the Northern Ritter Range pendant south of Tioga Road. These pendants contain continental-margin sediments that were later metamorphosed by heat from the Sierran granitic magmas.

Rocks of the Morrison block are considered transitional between the deep-marine rocks found to the northwest and the mostly shallow-water rocks exposed to the east. They were deposited along a passive continental margin; that is, there was no active mountain-building going on at the time. This all ended in the Late Triassic with the start of a spreading center in the Pacific and the consequent subduction and arc volcanism along the west coast.

The farther back in time you go, the harder it is to unravel exactly what took place. It is not clear whether the Paleozoic rocks in the eastern Sierra roof pendants were deformed by one or more than one episode of mountain-building. The earliest event might have been the Late Devonian–Early Mississippian Antler Orogeny, which involved mainly east-directed thrusting. This was followed by the Permian–Triassic **Morrison Orogeny**, elsewhere called the **Sonoma Orogeny**, about 250 million years ago, thought to represent a separate and much older subduction event and the associated accretion of a volcanic island arc onto the continent, accompanied by intense folding. All those who study this area seem to agree that thermal metamorphism occurred when Jurassic–Cretaceous Sierran granite intruded into the rocks, and that Cenozoic normal faulting uplifted the Sierra over the past 5–10 million years, tilting the rocks to the west.

Rocks of the Saddlebag Lake pendant, exposed at the Ellery Lake roadcut, are thus highly disrupted marine sedimentary rocks of ancient vintage. They were probably carried eastward by Devonian–Mississippian thrusting, perhaps metamorphosed by contact with nearby Devonian–Mississippian magmas, deformed again by Late Triassic subduction, heated again by Triassic–Cretaceous magmas, and then faulted, tilted, and uplifted during the Cenozoic. After all that these rocks have been through, it's a wonder that geologists can tell anything about them.

As I drive east from Ellery Lake, I notice that the road goes back into light-colored Sierran granite. While my wife drives, I notice that there are patches of dark metasedimentary rocks scattered along the road to Lee Vining. These are small, isolated roof pendants that foundered and sank into the rising granite magmas.

I also notice that the valley appears to be mostly U-shaped, as if carved by a glacier. The next stop confirms this observation.

To get to the Lateral Moraine stop, continue driving east on Tioga Road 8.7 miles (11 min) to a pullout on the right (south) side of road (37.939745, –119.121532). The lateral moraine is the ridge on the south side of the canyon.

STOP 28: LATERAL MORAINE

The topography starts to flatten and become gentler as we drive into the lower part of the canyon. A high, featureless, tree-covered ridge rises abruptly from Lee Vining Creek and extends eastward along the south side of the canyon. The first clue as to what's going on is at the Poole Power Plant turnoff that leads to the Moraine Campground. If you try to climb the steep, lightly wooded slope above the campground, you will find that it consists of poorly sorted boulders and

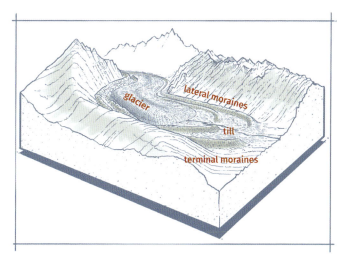

Lateral and terminal moraines (Adapted from Huber, 1987, p. 45)

other debris. The high ridge is a **lateral moraine** that follows Lee Vining Canyon and ends just before US-395.

Lateral moraines form along the margins of glaciers. This particular moraine was deposited along the edge of one of the short mountain glaciers moving east down the escarpment of the Sierra Nevada. Unlike glacial till, lateral moraines are ridges of coarse debris consisting of unsorted loose rock and sand plucked from the sides of the glacier, or carried to the fringe of the glacier by avalanches or tributary glaciers. The material is deposited when the glacier melts, and shows the greatest extent of the glacier.

Moraines provide clues to the glacial history of an area. Among these clues, a lateral moraine reveals the range of the glacier—how wide it was and how long it was. By studying the types of rocks and sediments in the moraine, geologists gain insights into the geology of the glacier's source area. Layers within the moraine represent different periods of glacial advance and retreat. Analyzing these layers can help reconstruct past glacial fluctuations and climate conditions. The moraine's size and shape also offer clues about the rate at which the glacier moved during its active phase. A larger, more extensive moraine usually indicates faster movement of the glacier, whereas a smaller moraine suggests slower movement. Dating the material found within the moraine and examining the moraine's relationship with other glacial features provide information about the timing and response of glaciers to climate changes.

Moraines are essentially an archive of a glacier's history and activity. They offer clues to the glacier's movement, its source area, and the changing environmental conditions at the time.

MONO LAKE AND LEE VINING

As I approach the east end of Lee Vining Canyon, there are glimpses of a broad desert valley stretching to the east, and an azure-blue lake that looks totally out of place in the arid landscape. The road just dropped 3,500 feet from Tioga Pass, passing from the alpine zone through pine forests to a pinyon-juniper-dominated landscape. Now I'm in a high desert where sagebrush and yellow-flowering rabbit brush thrive. Everything is much drier, if not hotter. The sweat in my eyebrows and armpits turns to a salty crust, and my lips start to crack. I wonder how there is a lake here.

The answer has to do with the last ice age. Earth has been subjected to multiple ice ages over the past 2 million years. During glacial intervals, the climate was much cooler and wetter; during interglacial periods, there was a lot of melting and runoff. Several large glacial lakes formed in the Great Basin due to the increased precipitation and glacial meltwater. For example, Great Salt Lake is all that remains of gigantic Glacial Lake Bonneville. Likewise, Mono Lake, elevation 6,378 feet, is the remnant of a large ice age lake, Glacial Lake Russell, which used to occupy a much greater part of this basin. Lake Russell formed around 760,000 years ago, making Mono Lake, its successor, one of the oldest continuously existing lakes in North America. Throughout its history, the lake averaged around 424 feet deep. The lake reached its greatest depth, 900 feet, at the end of the last ice age around 12,000 years ago. At that time, it extended several miles into Nevada. At its maximum, the lake surface reached 7,200 feet in elevation and would have covered about 338 square miles, compared to its meager 70 square miles now. Whenever the lake level was stable for a period of time, wind-generated waves would gnaw away at the shoreline, creating **wave-cut terraces**. One such prominent terrace surrounds the Mono Basin at 7,000 feet, suggesting a period of stable climate. Since the end of the last ice age, and despite being fed by streams flowing off the Sierra Nevada, Mono Lake has continuously shrunk due to evaporation in the parched climate that ensued. The lake has no outflow, and as a result, the lake water has become increasingly saline and enriched in carbonates. Today, it is about twice as salty as ocean water and is as alkaline as chlorine bleach. A white crust along the shore is a combination of salt and calcium carbonate, the mineral that makes **limestone**. The maximum water depth today is a mere 158 feet.

Stepped terraces mark the ancient shorelines on the mountainsides to the west and in the basin to the east. In satellite images, they look like successively smaller bathtub rings around the lake, revealing how the lake shrank as it dried up over the ages.

Even as late as 1941, the lake level was about 45 feet higher than it is today. In that year, a thirsty Los Angeles began diverting water from the lake by means of the LA Aqueduct. The lowering water level was accompanied by a further increase in salinity. In response to lowered lake levels and the threat to local wildlife, a group of concerned residents formed the Mono Lake Committee and fought the water diversion. In September 1994, the California State Water Resources Control Board voted to amend water licenses held by the Los Angeles Department of Water and Power to allow the lake level to rise a modest 17 feet. It may not sound like much, but the low water levels exposed a land bridge to

Negit Island, a major migratory bird nesting area. The land bridge allowed predators such as coyotes to cross over and devastate bird populations. The slightly higher water level prevented this.

The ecosystem of Mono Lake provides a nesting habitat for thirty-five species of shorebirds, and the brine shrimp that thrive in its waters feed over two million migratory birds each year. Mono Lake supports the second-largest breeding colony in the world of California gulls. Other birds that nest at Mono Lake include snowy plovers and Caspian terns. The presence of water and the abundance of brine shrimp make Mono Lake a reliable source of food for birds and a major migratory stopover for up to 750,000 eared grebes in an average season. The lake is a major stopping point for up to 80,000 Wilson's phalaropes and 65,000 red-necked phalaropes each year.

Hard on the shores of Mono Lake is the picturesque village of Lee Vining, population fifty-nine in 2020. The story of Lee Vining the town begins with Leroy Vining the man. A former Texas Ranger, Vining came to the Sierra to prospect for gold and started a mining camp here in 1852. It was called Poverty Flat because farmers couldn't grow anything. When his mining enterprise didn't pan out, he built a lumber mill that provided timbers to nearby mines. The poor guy died in 1863 when a pocket pistol he was carrying accidentally went off and severed an artery in his groin. Fast-forward sixty years: in a 1926 land-development scheme, Poverty Flats was subdivided into town lots and renamed Lakeview to entice buyers. There was little interest, and that name didn't stick either. The inhabitants renamed it again as Lee Vining in 1957. Today it serves largely as the eastern entrance to Yosemite National Park and headquarters for the Mono Basin Scenic Area Visitor Center.

I am now accompanied by my son and his wife. Before we leave this charming outpost, it is worth noting that fine dining is available. On our second night in the area, we stumble onto and eat at the Whoa Nellie Deli, an ingenious name that is too clever by half, in that it is not a delicatessen like any I've known. In fact, it is a diner attached to the Mobil gas station, and along with the simple but hearty fare, it is decked out with all the kitsch that tourists are expected to desire, from antifreeze to postcards, from cheap hats to T-shirts, and, maybe most important, twenty-four-hour restrooms. We have a hot dog, a burger, and vegetarian chili, buy a plush rainbow trout stuffed animal for the kid back home, and drive on.

Mono Lake is scenic, a deep-blue pool at the base of the Sierra escarpment. I find it most alluring in the spring, when the desert floor is blanketed in multihued wildflowers. The lake is a favorite of bird watchers, but also has interesting geology. There are **tufa** pinnacles just offshore from the Mono Lake Natural Reserve parking area (Stop 30), and good descriptions of the geology and landscape at our next stop, the visitor center.

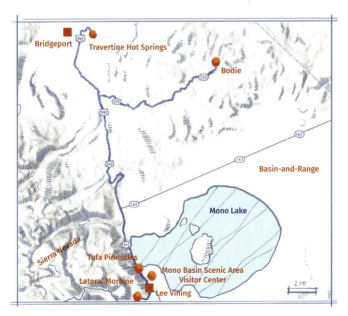

Map of the route from the lateral moraine to Mono Lake, Bodie Ghost Town, and Travertine Hot Springs

To get to the Mono Basin Scenic Area Visitor Center, continue east on CA-20 to US-395; turn left (north) on US-395 and drive a total of 2.4 miles (5 min) to the Mono Basin Scenic Area Visitor Center parking area at 1 Visitor Center Drive (37.966289, −119.120266) on the right (east) side of the highway.

STOP 29: MONO BASIN SCENIC AREA VISITOR CENTER

The visitor center sits on a slight rise with a scenic view of the lake. The building is modern, with natural light flooding in from its many windows. It has displays that explain the basin's geology, ecology, and human history. A twenty-minute film gives an introduction to the lake and basin. Photo and art exhibits showcase local landscapes. Rangers are available and eager to answer my questions. We wander the nature trails that lead from the center to the lake. The interpretive displays along the way explain various aspects of the ecosystem. For more information, see these websites: https://www.monocounty.org/listing/mono-basin-scenic-area-visitor-center/764/ and http://www.parks.ca.gov/?page_id=514.

Looking west, I see a sharp break between the mountains and the flats. This is the fault scarp, the Sierran Frontal Fault zone. A short distance up the scarp is the 7,000-foot-level of the ancient shoreline of Glacial Lake Russell. At its maximum, Lake Russell overflowed through the east side of Mono Basin, through the town of Benton, and eventually drained into the Owens Valley.

The light-colored rocks along the mountain front are Sierran granite; the darker and reddish-colored rocks are part of the Log Cabin Mine roof pendant. The Log Cabin Mine pendant, like the other pendants, consists of old continental-margin sediments that range in age from Ordovician to Mississippian. These rock formations are, from oldest to youngest, **slate**, quartzite, hornfels, black chert, and quartz sandstone, which indicate that the oldest rocks were deposited as muds in an ancient deep-marine environment. As the

Mono Lake as seen from the visitor center

sediments get younger, they indicate a shallowing marine environment, ending with a sandy beach setting. As with the Saddlebag Lake pendant that we passed higher up the eastern slope, these rocks too have been metamorphosed to varying extents.

To get to the tufa pinnacles at Mono Lake, continue north on US-395 to the Mono Lake Natural Reserve parking area on the right (37.976975, −119.133872), for a total of 1.6 miles (4 min).

STOP 30: TUFA PINNACLES AT MONO LAKE

From the visitor center, it's a short hop to the Mono Lake Natural Reserve parking area. This spot looks vaguely familiar, although I've never been here before. Then it hits me: Clint Eastwood's 1973 movie *High Plains Drifter* was filmed here. It's one of those movies that make you wonder why you're enjoying it so much when the plot pretty much sucks. In any case, the fictional mining town of Lago was set on the shore of Mono Lake, and the dramatic landscape is memorable.

From the Natural Reserve parking lot, we take the path toward the lake to see the tufa pinnacles. As we approach the lake, the first thing that hits us is the smell. I'm not sure what word describes it best: foul, noxious, putrid, or rank. Maybe rancid. It turns out that the stench is from dead brine shrimp washed up on the shore in their millions. Then I notice the flies. Alkali flies feed off the dead shrimp. Others, too, have remarked on this unexpected experience. Mark Twain does not sound favorably impressed when he commented on the lake and its critters in *Roughing It*:

> Mono Lake lies in a lifeless, treeless, hideous desert, eight thousand feet above the level of the sea, and is guarded by mountains two thousand feet higher, whose summits are always clothed in clouds. This solemn, silent, sailless sea—this lonely tenant of the loneliest spot on earth—is little graced with the picturesque. It is an unpretending expanse of grayish water, about a hundred miles in circumference, with two islands in its centre, mere upheavals of rent and scorched and blistered lava, snowed over with gray banks and drifts of pumice stone and ashes.... Its sluggish waters are so strong with alkali that if you only dip the most hopelessly soiled garment into them once or twice, and wring it out, it will be found as clean as if it had been through the ablest of washerwomen's hands....
>
> There are no fish in Mono Lake—no frogs, no snakes, no pollywogs—nothing, in fact, that goes to make life desirable. Millions of wild ducks and sea-gulls swim about the surface, but no living thing exists under the surface, except a white feathery sort of worm, one half an inch long, which looks like a bit of white thread frayed out at the sides. If you dip up a gallon of water, you will get about fifteen thousand of these. They give to the water a sort of grayish-white appearance. Then there is a fly,

which looks something like our house fly. These settle on the beach to eat the worms that wash ashore—and any time, you can see there a belt of flies an inch deep and six feet wide, and this belt extends clear around the lake.[1]

Much of Twain's humor, of course, comes from exaggeration, but here he hits pretty close to the mark. As I mentioned earlier, carbonates in the water make it highly alkaline, with a pH close to 9.8, similar to that of household glass cleaners.* Once you get past the smell and the flies, there are the peculiar columns of rock rising out of the lake. These delicate spires and knobs form striking and bizarre pinnacles that rise up to 30 feet above the lake surface, like the giant dripsand castles I used to make with my kids at the beach. The tufa pinnacles occur as individual mini-islands or as clusters of rock pillars.

    The light-gray to rusty-brown tufa rock that forms these odd-looking pinnacles is calcium carbonate. This rock precipitated out of the lake water when fresh groundwater, saturated with calcium, encountered the lake's carbonate-rich brine at underwater springs. These pinnacles, then, are made up of the same scale that builds up beneath a leaky faucet. They grew, layer on layer, beneath the surface of the lake. In fact, they could rise only until they encountered the lake surface, and thus give an indication of the water depth at the time they were forming. Sometimes there are broken pieces of the tufa scattered along the shore. The layering in the tufa was caused by algae growing on old surfaces, which were then covered by newly precipitated deposits, and so on over and over.

---

* The pH scale measures acidity and alkalinity. On the pH scale, a value of 0 is the strongest acid, 7 is neutral, and 14 is the strongest base. (Alkali and base refer to the same thing.) Liquid drain cleaner is about 14, seawater is around 8, human blood is 7.4, vinegar is around 3, and hydrochloric battery acid is 0.

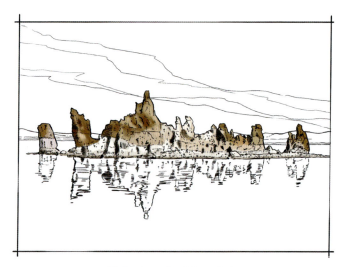

Tufa pinnacles, Mono Lake

Tufa pinnacles are exposed today because of the lowering lake level. These rock towers appear in old lakebeds throughout the Great Basin, anywhere groundwater springs occurred beneath lakes that have since dried up.

Mono Basin, the valley around the lake, was formed by faulting that began about 3–4 million years ago. This is the same faulting that ratcheted-up the east side of the Sierra Nevada. At the same time as the Sierra went up, the area to the east dropped down and began to extend in an east–west direction. The vertical uplift of the Sierra with respect to the valley bottom has been estimated at between half an inch and four feet per thousand years. This movement, spread over at least 5 million years, has caused more than 11,000 feet of offset here between the granite beneath the valley and the granite up on the Sierra Crest.

There are small volcanoes and **cinder cones** both around and in Mono Lake. Volcanism near Mono Lake is a result of faulting of the Earth's crust related to Basin-and-Range extension. Faults penetrating deep into the crust provide pathways for black basaltic lavas to push their way up to the surface. The lavas have covered large areas near here over the past 2.5–3 million years. The two islands in the lake, Negit and Paoha, are both volcanic, but these are light-colored lavas rich in silica. Negit is a 1,600-year-old **dacite** lava dome, and a small nearby basalt flow may be less than 220 years old. Paoha is composed of lake-bed sediments uplifted along a broad arch, and contains a rhyolite cinder cone that is only 250 years old. It would have erupted within the memory of the local Native people. Paoha lacks evidence of the 220-year-old shoreline at an elevation of 6,456 feet, suggesting that uplift of the island occurred between the late 1700s and late 1800s. And here is some interesting trivia: during World War II, Pahoa Island was used to simulate nuclear blasts using conventional explosives.[2] I'm not sure what the bomb scientists learned, but I'm pretty sure it wasn't good for migratory bird nesting.

Black Point is a small volcano on the north side of Mono Lake. The wave-cut shorelines scratched on its flanks indicate the falling lake levels. Black Point would have been under water when it erupted 13,000 years ago, a time when Glacial Lake Russell was full to the brim.

A short drive north of Mono Lake is a delightful hot springs and picturesque gold-mining ghost town. Both are related to and near the southernmost extent of the Cascade Range volcanic province. Other than the small tourist community of Bridgeport (population 550 in 2020) and old mines and prospect pits, there is not much more out here.

BODIE AND TRAVERTINE HOT SPRINGS

Bodie ghost town, which is all that remains of a once-thriving gold-mining district, and the hot springs near Bridgeport are tied together by their geologic history. Both are part of the southern Cascade volcanic arc. The gold was found in veins deposited by hot springs, and the hot springs are a vestige of the 8-million-year-old volcanic field that forms the hills.

To get to the Bodie ghost town from the tufa pinnacles, continue north on US-395 for 17 miles to CA-270; turn right (east) onto CA-270 and drive to the Bodie parking area (38.213513, −119.015540), for a total of 29.8 miles (42 min).

STOP 31: BODIE STATE HISTORIC PARK AND GHOST TOWN

Bleak is the word that comes to mind when describing the Bodie Mining District. The bone-dry hills are dotted with volcanic cones, **volcanic craters**, lava domes, lava flows, fumaroles, and hot-spring deposits, and there are mine dumps next to prospect pits. It is high desert, with rocky and thin soil and low-growing silver sagebrush adding faded blue-green to the gray-brown background. There are no trees to be seen in the barren landscape. I am soon to learn why.

Dust rises behind our vehicle as I come to a stop for a flock of sheep being driven across the gravel road. I wonder what they find to eat out here. The ghost town doesn't look like much, just a bunch of ramshackle wood structures and a large stamp mill rusting on the hillside. It looks like a good wind would knock it all down.

Bodie ghost town, at 8,379 feet elevation, is remarkably well preserved. Over 170 buildings still stand in this state park. We roam the town, peeking into windows of wood homes that lean like drunks and checking out the musty old schoolhouse, the saloon, and the general store. The cemetery is a curious mix of history and humanity: here lie the men, women, and even children who came to this desolate spot to carve out their own piece of the American dream. How many stories of struggle and success, or heartache and ruin, are buried with them?

No one did want to live here until some prospectors, discouraged by the dwindling pickings in the goldfields of the Sacramento Valley, came looking for new gold in the late 1850s. In 1859, four prospectors discovered gold nuggets in dry riverbed gravels high in the hills east of town. They mined the placer deposit until an unexpected November blizzard killed one of them, Bill Bodey. According to local lore, a sign painter in nearby Aurora, Nevada, misspelled the name when he lettered a sign for Bodie Stables. The name stuck, and by 1862 the mining district, populated by about twenty miners, was known as Bodie.

Bleak terrain and poor returns discouraged much interest in the district. There were scattered dry placer deposits and a few tunnels into low-grade quartz veins until 1864. In that year, the Bunker Hill Mine had a roof collapse that exposed a rich vein of gold. That got everyone's attention and brought more miners to these hills.

In 1876, the Standard Mining Company discovered a profitable deposit of gold-bearing ore that turned Bodie into a Wild West boomtown and attracted San Francisco speculators. Their gamble on Standard Mining paid off when it produced $784,523 in gold and silver during 1877. The fantastic yields prompted Standard Mining to build a stamp mill in mid-1877. The mill burned down in 1898, but was replaced by the mill that survives today. The Bodie Mining Company discovered two rich veins in 1878. In that year, the syndicate mill crushed 1,000 tons of ore from the Bodie Mine in one month, yielding a whopping $601,103 in gold.

The rich strike in the Bodie Mine was followed by the discovery of the incredibly productive Fortuna Lode and the vast Main Standard Ledge. Everyone was convinced that Bodie would be the next Comstock bonanza. By late 1878, there were twenty-two mines operating in the district.

Thousands of fortune seekers poured in to Bodie, making it one of the West's wildest towns. Shootings were commonplace, and Bodie got a reputation for frontier violence that rivaled Tombstone, Deadwood, and Dodge City. "Saloons and gambling halls abound," reported San Francisco's *Daily Alta California* in June 1879. "There are at least sixty saloons in the place and not a single church."[3] At its peak, the population of the scruffy, ramshackle town was estimated to be between eight and ten thousand individuals. Saloons, dance halls, opium dens, and brothels lined the streets. There were sixty-five saloons along the mile-long Main Street alone. The Reverend F. M. Warrington visited Bodie in 1881 and described it as "a sea of sin, lashed by the tempests of lust and passion."[4]

Local legend tells of a little girl who was moving from San Francisco to Bodie with her family. She reportedly wrote in her diary, "Goodbye, God: we are going to Bodie." An irritated Bodie editor responded that the girl had been misquoted. What she had really said was "GOOD. By God, we are going to Bodie."[5]

The estimated 60 miles of tunnels in the district all needed to be shored up with timber. Consequently, all the trees were cut for miles around, leaving the hills even more barren. And trees don't readily grow back in an arid climate. By 1880, many of the mines were played out or were just not economical, and miners and their families began to head elsewhere. Bodie's deepest shaft, in the Standard Mine, reached the 1,200-foot level in April 1882. Only a few mines were still producing gold when, in 1890, cyanide heap leaching revolutionized gold and silver production. In this process, dilute sodium cyanide is used to dissolve precious metals. The crushed ore is placed on a waterproof liner, then the cyanide solution is sprayed or dripped over the ore, dissolving any gold and silver in the solution. The "pregnant" solution containing dissolved precious metals continues percolating through the crushed ore until it reaches the liner at the bottom of the heap, where it is then collected in a pond or other container. The most common method of removing gold from the solution is to use activated charcoal to absorb any gold and silver, producing doré gold (a gold-silver amalgam, or mixture). Otherwise, zinc powder can be added to the solution to cause precipitation of a gold-zinc sludge that is then refined elsewhere.

This method was an improvement over the mercury amalgamation process previously used by the stamp mills. In that process, crushed ore was washed over mercury-coated copper sheets, and fine gold particles would glom onto and form an amalgam with the mercury. The amalgam was scraped off and the gold then separated from the amalgam by heating and evaporating the mercury. It is not clear whether the miners were aware of just how lethal these methods were at a time before there were respirators. I doubt anyone working in the gold-refining business lived to a ripe old age. Regardless, the cyanide process led to a mini-boom where a lot of the old mine **tailings** were reprocessed using the new method. There is no record of how many died from mercury and cyanide poisoning.

For three more decades, a handful of mines supported a population of about eight hundred, mostly miners and their families. The Bodie Company, its ore bodies depleted, sold out in 1896 to the Standard Mining Company, which lasted until 1913. The Standard was Bodie's most profitable mine, producing slightly more than $18 million in gold and silver over its thirty-seven-year life.

Treadwell-Yukon, part of a multinational mining conglomerate, reopened the Red Cloud Mine in 1929. The company suffered heavy losses and left after two years. The mining district and the town it once supported were nearly abandoned when fire engulfed the wooden town in 1932. The last mine closed in 1942, due to a War Production Board order shutting down all nonessential mines during World War II. Over the years 1860–1941, Bodie's mines produced approximately 1.5 million ounces of gold and more than 1 million ounces of silver. Bodie is considered the most productive mining district in the Basin-and-Range province. The town was completely deserted by 1950.

Bodie's surviving structures were taken over in 1962 by California's Department of Parks and Recreation and preserved in a "state of arrested decay." This meant that the natural decay of the buildings would be prevented to the extent possible, but no reconstruction would be done. Bodie

Bodie stamp mill, homes, scattered mine dumps, and barren hills

is considered the West's largest, best-preserved ghost town. The town, now Bodie State Historic Park, is a state Registered Historical Landmark.

My only disappointment is that there is no access to the mine dumps. I have spent days poking through ore minerals in old mine dumps scattered across the country. I don't know the reason these dumps are fenced off, but it may have to do with traces of cyanide left over from the ore extraction process. In that case, it's probably a good idea.

The gold in the Bodie Mining District is related to the underground source of volcanism and a series of faults that allowed a hydrothermal (hot-water) system to develop. Groundwater, heated by magma and containing dissolved metals, forced its way to the surface along faults, where it deposited silver and gold in milky-white quartz veins in a **volcanic plug**. Volcanic plugs are the vents, or throats, of the old volcanoes, the jumble of lava and ash just below the crater.

The basement rock, or bedrock, under the district is 150-million-year-old Sierra Nevada granite that intruded into even older **gneiss** and **schist**. Gneiss is an attractive banded stone that was formerly some other rock, sedimentary or igneous, but was heated almost to the point of melting. Schist is also metamorphic, derived from mud or clay that has been heated under great pressure. Above these rocks are 5–13-million-year-old volcanic **tuffs**, lava flows, and plugs. The vast majority of gold was produced out of an 8-million-year-old plug around Bodie Bluff and Standard Hill. The Miocene volcanics, up to 1,200 feet thick, erupted from numerous local **stratovolcanoes**, those classic cone-shaped volcanoes that look like Mount Fuji. Dacite tuff **breccia**, a medium-dark busted-up volcanic ash, is the most common rock type. The richest gold-bearing quartz veins run along north–northeast faults. The best quartz veins also contain silver, **electrum** (a gold-silver alloy), pyrite (fool's gold), **galena** (lead ore), **sphalerite** (zinc ore), **chalcopyrite** (copper ore), and other minerals. The veins can be anywhere from several inches to as much as 90 feet wide, and run for thousands of feet across the countryside. Most of the veins extend downward no more than 500 feet, but a few go as deep as 1,000 feet.

The chemistry of the ore minerals at Bodie indicates an ancient hot-spring system that was active 7–8 million years ago. Precious metals and minerals were dissolved in heated groundwater similar to present-day hot springs in the Bodie Hills. The hot water flowed through the broken rock in fault zones, and the ore minerals then precipitated in quartz veins and in breccias—zones of broken rock fragments—as the temperatures and pressures decreased. The key to these deposits, then, was hot water. And that takes us to our next stop, Travertine Hot Springs on the west side of the Bodie Hills.

To get to Travertine Hot Springs, return west to US-395 and turn right (north); after 19 miles, turn right (east) onto Jack Sawyer Road (the sign says County Animal Shelter); follow the signs to Travertine Hot Springs parking area (38.245700, −119.204072), for a total of 20.3 miles (37 min).

STOP 32: TRAVERTINE HOT SPRINGS

We go looking for Travertine Hot Springs on a cold winter day. The hot springs are at the end of a short gravel road about two miles southeast of Bridgeport. The first thing I see is bare spots where the sagebrush gave way to white crusty soil. Then I notice steam rising into the clear morning air.

Several large horizontal terraces, painted yellow and green from the algae living there, face west toward the Sierra Nevada. These mineral mounds are **travertine**, a type of limestone that precipitates when the hot, mineral-laden groundwater emerges and starts to cool. They are like the tufa pinnacles at Mono Lake, except they are related to hot springs and didn't form under water.

It is a scenic setting: the low rolling hills are covered in fragrant Utah juniper and pinyon pine, Great Basin sage, and antelope bush. The air is filled with the calls of Clark's nutcrackers (chunky gray birds with black wings); mountain bluebirds; and the harsh, chattering *wock, wock-a-wock* of yellow-billed magpies. This area of high desert gets quite hot in the summer and very cold in winter. The springs are popular with the locals year-round, but for me there is something special about sitting in a steaming-hot pool of water when there is snow on the ground. The hard part is leaving the pool, because the air is freezing.

There are few visitors when we arrive at these undeveloped springs on a midwinter day. The springs do get a fair amount of tourist traffic, though, and there have been complaints about visitors trashing the site. We all would do well to respect these pools and their environment.

Over the years, bathers have built a series of rock pools down the hillside. The springs emerge at 180°F from fissures in large travertine mounds. The water cools to 115°F as it moves down through the pools.

The hot mineral springs have long been used by the indigenous Paiutes, who treat this as a holy place. The early settlers enjoyed the waters too. And yet, because there is

Two of the hot-spring pools and an algae-colored travertine mound

always someone looking to make a buck, 60 tons of the hot-spring travertine was mined in the 1890s and used to build the interior facings of San Francisco City Hall. The rock also became a popular fireplace stone in Southern California. More recently, energy companies tried to tap into the geothermal heat, but after a few holes were drilled, the idea was abandoned because of the soft nature of the rock—and in response to local opposition.

The springs flow from broken rock at the intersection of two faults. The faults cut the Miocene–Pliocene Willow Springs Formation, an 8-million-year-old andesite lava exposed in the western part of the Bodie Hills Volcanic Field. The volcanics are considered part of the ancestral southern Cascade volcanic province.

After a relaxing dip in the hot pool, we climb into the car, but it won't start. The battery is dead, like, not even a click or a woo-woo-woo from the engine. I pop the hood and open the battery cover; the posts and contacts are just covered in that white powdery corrosion residue. I must look pretty dejected, because the couple next to us ask if we need help. We don't have jumper cables, but another family with a large pickup truck and lots of kids does. They get us going on the fifth try. To paraphrase a favorite saying, I've always been

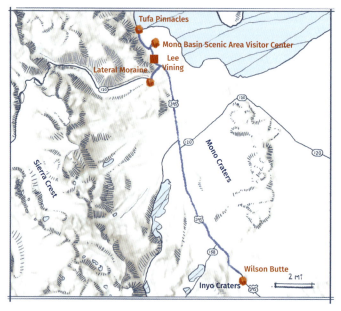

Map of the route from Mono Lake to Wilson Butte

saved by the kindness of strangers. I try to call the auto club and arrange for them to meet us in Bridgeport to replace the battery, but they don't consider this to be "roadside assistance." So we drive straight to town and park at a garage, since it is late and the place is closed, and we walk to our hotel. Lo and behold, the car starts the next morning. But I'm not taking any chances: we get a new battery and head south.

As we pass Mono Lake again, we are about to cross a transition from the ancestral Cascade volcanics to the much more recent volcanism of the Long Valley Caldera. It is easy to see, too. The rounded, rolling old hills give way to sharp pimples and bumps of recent eruptions. The southern Cascade volcanics and the Sierra Nevada were both a result of subduction of the oceanic plate beneath North America over the last 150 million years, give or take a few million. The Long Valley Caldera, on the other hand, is a result of Basin-and-Range extension that tapped into near-surface magma in the past million years, and the caldera is still very much alive and active.

To get from Travertine Hot Springs to Wilson Butte in the Inyo Craters, return to US-395 and turn left (south); drive south for 40 miles (47 min) and pull over on the right shoulder just south of Wilson Butte (37.780297, −119.020832).

THE LONG VALLEY CALDERA AND SIERRA ESCARPMENT

The drive south from Lee Vining is new to me. This is California's eastern corridor, its unpopulated hinterland, the mirror opposite of the 50-mile-wide coastal strip where 80 percent of the state's thirty-nine million inhabitants live.

A long grade climbs out of the Mono Basin, and we enter rolling hills cloaked in sparse pine forest. The highway climbs again near Crestview. This ridge marks the northern rim of the Long Valley Caldera.

Growing up in Southern California, I knew there were volcanoes in the northern part of the state. Mount Shasta and Mount Lassen are both classic stratovolcanoes, and Mount Lassen erupted only a hundred years ago, between 1914 and 1917. But I was totally unaware of the little-known caldera system near the ski-resort town of Mammoth Lakes. A **caldera** is the crater left behind after a really large volcanic eruption, where so much lava and ash are spewed into the air that the magma chamber collapses in on itself. The collapse of the Long Valley Caldera caused an even larger secondary eruption that blanketed thousands of square miles of the western United States with volcanic ash.

Most Californians aren't aware that a huge volcano exploded in northeastern California 760,000 years ago, and again 740,000 years ago. The gigantic eruptions blew out 140 cubic miles of rock and pyroclastic flows. That's equivalent to an area 10 miles by 14 miles covered by a mile-thick carpet of ash. Imagine glowing clouds of hot ash, like what buried the

Roman towns of Pompeii and Herculaneum in 79 AD. That is what a pyroclastic flow is. The superhot ash and volcanic gases first incinerate, then bury everything in their path. For those who were not around in 1980, an earthquake at 8:32 a.m. on Sunday, May 18, caused the entire north face of Mount St. Helens in Oregon to slide away, the largest landslide in recorded history. This had the effect of uncorking the entire mountain. Partly molten rock, rich in high-pressure gas and steam, suddenly exploded in a hot mix of lava, dust, and broken rock. Harry Truman, owner and caretaker of the Mount St. Helens Lodge at Spirit Lake on the north flank of the mountain, was killed when a pyroclastic flow buried his cabin under 150 feet of volcanic debris. The eruption released a column of ash that climbed 80,000 feet (15 miles) into the atmosphere, turning day into night, and the prevailing winds carried the ash over eleven US states and multiple Canadian provinces. That was Mount St. Helens. The Long Valley eruptions were 2,000 times larger than the 1980 eruption of Mount St. Helens.

The Long Valley Caldera is the residue of that ancient and very large volcano. It is 10 miles wide by 20 miles long. Long Valley is bounded by faults that define the semicircular caldera, and it is filled with between 5,000 and 10,000 feet of volcanic debris. The town of Mammoth Lakes sits inside the caldera at the southwest corner. Ash and pulverized rock from the last eruption, the Bishop Tuff, covers about 850 square miles of eastern California with up to 660 feet of ash. Thinner layers of ash as far away as Nebraska have been chemically identified as Bishop Tuff, showing the extraordinary reach of this eruption.

Like the Mono Basin, this area is still volcanically active. The youngest eruption, at Mammoth Knolls along the caldera rim, is 100,000 years old, but hot springs, fumaroles, and earthquakes that are active today are the result of magma moving near the surface. The magma is so near the surface that groundwater at only 600 feet depth has been heated to

Basin-and-Range 175

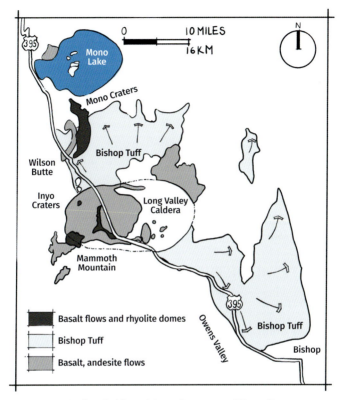

Long Valley Caldera, Mono Craters, and Inyo Craters. The oldest formations are 3.6-million-year-old basaltic and andesitic lava flows. The Bishop Tuff is 740,000 years old. Mammoth Mountain, Mono and Inyo Craters, and local basalt flows are less than 110,000 years old. (Map adapted from Hill et al., 2000)

340°F, well above the boiling point. The heat flowing from this magma is high enough that three power plants are able to tap into the geothermal energy to produce about 40 MW of electricity every year, enough energy to power nine thousand homes for 25 years.

The area has been **seismically active** since a magnitude 5.4 earthquake and a swarm of aftershocks occurred there in the years 1978 to 1980. The mostly small earthquakes and surface movements are constantly monitored by the US Geological Survey and the California Volcano Observatory. The 1980 quakes domed the area upward by two feet, leading the US Geological Survey to issue an eruption alert. The activity was thought to be evidence of magma movement in a chamber below Mammoth Lakes. The eruption didn't happen (that time), but the alert caused the local economy, especially the real estate market, to suffer. As at the fictional Amity Island in the movie *Jaws*, Mammoth Lakes businesses did not appreciate the bad publicity. The eruption alert had an adverse impact on geologists, too. Following the alert, signs on café doors read "Geologists not welcome."

STOP 33: WILSON BUTTE, INYO CRATERS

We're driving south from Lee Vining. The first gray volcanic hills to the south are Mono Craters. The "craters" are actually **rhyolite domes** aligned along a curving system of faults. The semicircular fault pattern suggests to some geologists that this is a caldera rim south of Mono Lake and that a volcanic center may exist beneath the Mono Basin. The domes along the highway here range in age from 600 to 35,000 years old.

Mono Craters extend south from Mono Lake on the east side of US-395, and Inyo Craters are west of US-395. Inyo Craters are a string of very young **obsidian domes** (a glassy rhyolite) and cinder cones between Mono Lake and Mammoth Lakes. The obsidian domes are aligned north to south, suggesting that they erupted along a fault or fissure. Some of the cinder cones are a mere 400–700 years old. The craters are a result of violent steam explosions. As magma rose along the fault, it came into contact with groundwater that then flashed into steam. The sudden expansion of water to steam created an explosion that blasted out a crater and deposited rock fragments (**tephra**) around the crater. Then the magma, a rhyolite, erupted into the crater and filled it. Rhyolite is the stickiest, most viscous type of

Wilson Butte, a bit over a thousand years old, as seen from the side of US-395

lava. Because of this it never flows far, and forms steep-sided mounds, or lava domes. If the eruptions continue, a rhyolite or obsidian dome spills over the crater and tephra, kind of like a natural muffin top.

Wilson Butte, northernmost of the Inyo Craters, is one such rhyolite dome. It erupted fairly recently, somewhere between 1,350 and 1,200 years ago, about the time the Vikings were beginning their rampages in Europe. The eruption produced pyroclastic ash flows that extended 5.5 miles and covered 23 square miles. This relatively small lava dome rises abruptly 400 feet above the valley floor. Wilson Butte is located just outside the north edge of the Long Valley Caldera.

Abundant hot springs and fumaroles in the area indicate that the region is still volcanically active.

To get from Wilson Butte to the Sierra Nevada Fault, drive 11.3 miles (10 min) south on US-395 to the Mammoth Lakes exit and pull over on the right shoulder of the off-ramp (37.642589, −118.919961).

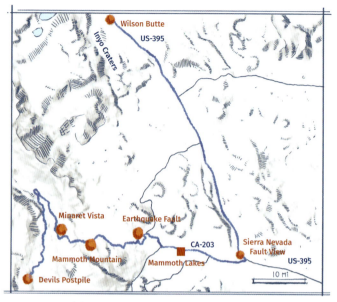

Map of the route from Wilson Butte to Devils Postpile

STOP 34: SIERRA NEVADA FAULT VIEW

Driving south from Wilson Butte, we reach the turnoff to Mammoth Lakes and pull over. The peaks are getting higher as we drive south. This stop provides a panoramic view of the eastern Sierra Nevada escarpment, an impressive granite rampart with abrupt cliffs along the north and east faces of the peaks. Between Mammoth Lakes and Bishop, 35 miles southeast of here, the fault has raised the Sierra up by 15,000 feet in the last 5–10 million years. Due to being partly buried under the valley and to erosion along the Sierra Crest, the current fault scarp only has about 4,300 feet of relief. In the foreground, I can see huge **alluvial fans**, triangle-shaped piles of gravel and sand shed off the uplifting range, emerging from short, steep valleys all along the base of the mountains.

To the casual observer, the Sierra Nevada Fault appears to be one massive break along the east flank of the mountain range. In fact, the fault is a series of near-vertical segments that overlap and line up along the edge of the mountains. The

View west to the Sierra escarpment from just south of Mammoth Lakes. A line of aspen turning fall colors flows like a river of molten gold down the alluvial fans.

offset along each fault segment is mostly vertical, with the west side up and the east side down, but there is evidence for some minor sideways offset as well. The sideways offset, like the offset on the San Andreas system, has the west side moving north and the east side moving south. The Sierra Nevada Fault zone extends almost 360 miles from the Garlock Fault and Indian Wells in the south to the Cascade Range near Mount Lassen in the north. Together with the nearby Owens Valley Fault, the Sierra Nevada Fault is responsible for uplifting the Sierra Nevada. But that begs the question: What was here *before* the Sierra Nevada was uplifted?

MAMMOTH MOUNTAIN AND DEVILS POSTPILE NATIONAL MONUMENT

Not that long ago, geologically speaking, the San Joaquin River flowed out of central Nevada, across the Mammoth Lakes area, and into the Central Valley of California. Then, maybe 5 million years ago, the Sierra Nevada began to rise, levered upward by faulting along the east side of the range. Lava flowed up along local faults in the river valley near Mammoth Lakes. The uplift and lavas severed the river from its eastern source area. Today, the San Joaquin River is confined to the western side of the Sierra. The Great Basin, including most of Nevada and much of western Utah, is totally cut off from the sea.

If you will indulge me, this presents an opportunity to digress and speculate. As Mark Twain said, "There is something fascinating about science. One gets such wholesale returns of conjecture out of a trifling investment of fact."[6] One of the great mysteries of the Grand Canyon is how the Colorado River cut the canyon into and across the Kaibab Uplift, a large uplifted block of continental crust. The consensus of most who have studied it is that there were *two* ancestral Colorado Rivers, one that flowed from the area around Las Vegas to the Gulf of California, and another that flowed *northeast* from the highlands near today's Grand Canyon toward New Mexico and southeast Utah. The tributaries to the Colorado on the east side of the Grand Canyon all flow northeast, suggesting a northeast-flowing Colorado River. Problem is, no one knows where this river would have flowed *to*. There are no Grand Canyon–derived gravels or river sediments to be found northeast of the canyon. In the event, between 5 and 6 million years ago the lower Colorado River eroded upstream, a process known as headward erosion. Headward erosion occurs when moving water erodes and deepens its valley opposite the direction of flow. The erosion usually occurs at the steepest part of the valley, often at a waterfall. The flow of water undercuts the steep section, causing the waterfall to migrate upstream. In the case of the Colorado River, the river began cutting into the Kaibab Uplift and captured the northeast-flowing upper Colorado River in a transaction known as **stream piracy**. We know this from the age and composition of Colorado River gravels found around Lake Mead. From that time onward, the now complete river flowed west and south and emptied into the Gulf of California.

So, what does this have to do with Mammoth Lakes and the headwaters of the San Joaquin River? My personal conjecture is that the pre-6-million-year-old upper Colorado River flowed north and east along its present course to around Grand Junction, Colorado, whereupon it turned hard left (west) and found its way across Utah and Nevada to the ancestral San Joaquin (or Yuba) River and thence to the Pacific through California. No, I have not done the research necessary to prove any of this. I just thought I'd throw it out there for your consideration.

To get to Earthquake Fault, continue on the off-ramp to CA-203 and turn right (west); drive toward Mammoth Lakes; turn right (north) on CA-203/Minaret Road and drive to the Earthquake Fault parking area on the right (37.653275, −119.000589), for a total of 5.9 miles (11 min).

STOP 35: EARTHQUAKE FAULT

We take CA-203 west through the town of Mammoth Lakes to the "Earthquake Fault" parking area. It is only a short walk to this gash in the volcanic landscape.

Earthquake Fault is a misnomer. Local boosters like to come up with catchy names for their scenic stops. This stop is not a fault, nor was it caused by an earthquake. A fault is a break in the rocks where the two sides slide past each other either vertically or horizontally. Technically, what I'm seeing here is a fissure. It is about 10 feet wide and 60 feet deep, and runs north–south toward Earthquake Dome. The sides of the rock fit together perfectly, like pieces of a puzzle, which means the crack opened up as the sides moved away from each other. The fissure opened around 550–650 years ago, during a time of volcanic activity that included eruptions at Inyo Craters and Deadman, Obsidian, and Glass Creek

Domes. This type of fissure forms when volcanic uplift, or bulging, causes fracturing and extension in rocks at the surface. Kind of like if you drew two adjacent parallel lines on a balloon before blowing into it; then, as you inflate it, the lines move apart on the expanding surface.

Because the bottom of this crack is in perpetual shade, and the area gets plenty of snow in winter, there is almost always ice in the cleft year-round. Early pioneers collected ice from the fissure in the summer. There is still snow in it when we visit.

This is a nice place to take a break. The tall pine forest has the fragrant smell of pine needles toasting in the sun. There are picnic tables, restrooms, and easy walking trails.

To get from Earthquake Fault to Mammoth Mountain, drive 2.5 miles (5 min) west on CA-203 to the Mammoth Mountain Ski Area Adventure Center parking area (37.651319, −119.036976). This is where you board the mandatory shuttle to get to Devils Postpile during the high season.

STOP 36: MAMMOTH MOUNTAIN

It is hard to miss this ski area. Mammoth Mountain and the ski resort it is famous for dominate the southwest rim of the Long Valley Caldera. The 11,059-foot-high mountain is a volcano formed by eruptions of silica-rich lavas that began about 400,000 years ago and lasted until about 57,000 years ago. It is considered dormant, or sleeping, but that is a subjective term. Ongoing earthquakes and gas and steam vents suggest that magma is moving not that far beneath the surface. In the 1990s, trees began to die on the mountain. The Forest Service determined that they were being killed by high levels of carbon dioxide (CO_2) in the soil. The source of the CO_2 is thought to be limestone (calcium carbonate) in the subsurface that is being heated by contact with magma. The heated carbonate gives off CO_2. The gas moves to the surface along faults, mainly on the south side of the mountain near Horseshoe Lake. The US Geological Survey monitors the mountain and found that it emits about 1,300 tons of CO_2 every day. And it is not only killing trees. In April 2006, on the slope of Mammoth Mountain ski area, three members of the Mammoth Mountain ski patrol suffocated when they fell into a snow cave above a fumarole on the slopes. Pockets of the heavier-than-air CO_2, known as *mazuku* ("evil wind" in Swahili), can pool in low spots, and have killed people and animals. Venting steam causes a semipermanent cloud over the area each winter.

The town of Mammoth Lakes is a thriving ski resort inside the Long Valley Caldera and next to Mammoth Mountain. The bad news, according to the US Geological Survey, is that it is nearly certain that the mountain will erupt again. The good news is that there is less than a 1 percent chance of an eruption in any given year. Just to be safe, the Mammoth Scenic Loop/Hwy 203, despite its name, was built to provide an additional escape route for the residents of Mammoth Lakes in case of an eruption.

Take the shuttle from the Mammoth Mountain Ski Area Adventure Center to Minaret Vista. If the shuttle is not running, from Mammoth Mountain continue driving 1.6 miles (5 min) west on CA-203/Minaret Road to the Minaret Vista turnoff on the right (37.656573, −119.061279).

STOP 37: MINARET VISTA

We continue west past the upscale Mammoth Mountain ski resort and its spider web of ski runs that cover the mountain. During high season this is as far as you can drive: a mandatory shuttle takes visitors the rest of the way to Devils Postpile. In the winter the road is closed due to snow, and access is by snowmobile, snowshoe, or skis. We're here in early fall, when the road is open.

Just above the saddle on San Joaquin Ridge, where the road begins its steep, winding descent into Reds Meadow, there is a parking area and vista point offering one of the best views in the entire Sierra. Across the valley to the west rise the imposing spires of the serrated Minarets and the large massifs of Mount Ritter and Mount Banner. Interpretive displays help to identify individual peaks. A number of trails begin here and lead up the ridge and down into the valley.

Minaret Vista provides a panoramic view of the Sierra Crest and the present-day headwaters of the San Joaquin River. The San Joaquin River today rises in the deep, smoothly glaciated

Minarets Crest from below Mt. Elektra
(*Woodcut © Tom Killion. Reproduced with permission.*)

valley in the foreground. The Minarets were named in 1868 by the California Geographical Survey, which reported: "To the south of Mount Ritter are some grand pinnacles of granite, very lofty and apparently inaccessible, to which we gave the name of 'the Minarets.'"[7] But only the lower part of the range is granite. The pinnacles and massifs are mostly Jurassic metavolcanic rocks that weathered to scenic crags. They are part of the Ritter Range roof pendant, up to 30,000 feet of metamorphosed volcanics that accumulated at the surface at the same time as the Sierra Nevada granites were working their way up from below.

In contrast to the approximately 167–174-million-year-old Jurassic volcanics across the valley, the relatively recent volcanics at our next stop have a curious twist that makes them unusually attractive.

To get to Devils Postpile, take the mandatory shuttle bus that leaves from the Mammoth Mountain Ski Resort Adventure Center. At off-peak times of year, you can drive the 7.4 miles (22 min) to Devils Postpile National Monument (37.629880, −119.084661). The road is closed in winter due to snow.

STOP 38: DEVILS POSTPILE NATIONAL MONUMENT

As a young man, I had hiked the John Muir Trail, which goes near Devils Postpile. I got as close as Reds Meadow, a mile or so away, but at the time I was more interested in a resupply and shower than seeing some volcanics. In the many years since then, this spot had been on my "to see" list, but, as the song says, life got in the way. So when I finally do get here, it is special. I mean, the stately columns forming the cliff like some gigantic geologic Parthenon are spectacular in and of themselves. But knowing something about how these natural columns form, and how glaciers polish rock, makes the experience even more meaningful.

This national monument is on the Middle Fork San Joaquin River. A short walk from the parking area brings visitors to the "postpile," a spectacular cliff of columnar-jointed basalt. The basalt flow erupted about 82,000 years ago, a geological blink of an eye. The location of the vent or fissure has never been found, but is thought to be in Reds Meadow near Upper Soda Springs.

Basalt lava has a lower silica content and is typically hotter than other types of lava. Because of this, it tends to have a lower viscosity and flows more readily than other lavas. Think of the difference between warm maple syrup (basalt) and cold honey (rhyolite). The basalt lava flowed down the valley like a red-hot river until it was blocked by a natural dam, probably a glacial moraine. The molten rock filled the valley behind this dam, creating a lava lake 400–600 feet deep.

As the flow slowly cooled, it formed these vertical cooling joints. As lava cools, it contracts, just as mud shrinks and cracks when it dries out. This type of jointing is fairly common in basalt flows, but Devils Postpile is an unusually well exposed section of very long, perfect columns.

Hexagonal columns of basalt at Devils Postpile

The cooling joints formed columns that are mostly hexagons (six sided) and pentagons (five sided). The almost perfect geometric shapes are a geological curiosity. These patterns are thought to be the geometries that form using the least amount of work—that is, that require the smallest force to crack the rock. This is the same type of cooling joints I saw from afar at the Palisades of the Sonoma Volcanics in Napa Valley. Only here I can walk right up to the base of the columns and climb onto the polished tops of the columns. Which I am only too happy to do.

Reds Meadow Valley has been subjected to several glacial episodes. The Tioga-stage glacier that occupied the valley around 10,000–20,000 years ago removed much of the upper flow and polished the surface. The top of the flow has a mirror-like polish, as well as **striations** (scratches) caused by the grinding movement of the heavy, grit-laden ice.

The deep-going fault or fissure that allowed the eruption of the Devils Postpile basalt is most likely one strand of the

Polished and striated hexagons and pentagons, top of the flow at Devils Postpile

eastern Sierra Nevada Fault zone that separates the Sierra from the Owens Valley.

OWENS VALLEY

On leaving Mammoth Lakes, I enter the Owens Valley. Owens Valley is the arid, flat-bottomed, north–south valley of the Owens River. It is located east of the Sierra Nevada and west of the White and Inyo Mountains. Ninety miles long and averaging 18 miles wide, it extends from Mammoth Lakes in the north to Owens Lake in the south. The valley lies within the Mojave Desert, along the western edge of the Great Basin. It seems incongruous that the indigenous Paiute name for the mostly arid valley, Payahüünadü, means "place of flowing water." However, wherever there *was* water in the barren Mojave Desert, the Paiute developed irrigation systems to grow food. Near the modern town of Bishop, the Paiute used irrigation ditches to grow plants such as spike rush, corn, squash, melons, gourds, and sunflowers.[8]

The eastern part of the state, and the Mojave Desert in particular, is a geologist's paradise. There is little besides rock and dirt, and thus few distractions. But if you time it just right, say a week or two after a rainstorm in early spring, the desert undergoes a marvelous transformation. The baked and blasted landscape is clothed in gold of another kind, golden poppies. The yellow-orange superbloom blankets the desert floor. This humble weed had earned the right to be the state flower of the Golden State.

During Precambrian and part of Paleozoic time, the area occupied by the Owens Valley was offshore from the western margin of what would later become North America. During middle and late Paleozoic time, the Antler and Sonoma Orogenies folded and pushed the mainly marine sediments eastward on thrust faults.

The Owens Valley is partly underlain by the Jurassic–Cretaceous-age Sierra Nevada regional granodiorite mass. During the Cenozoic, the area was subject to Basin-and-

Range crustal extension. The Owens Valley is the westernmost basin in the Basin-and-Range geologic province: it forms the boundary between the Basin-and-Range and the Sierra Nevada. It is one of the youngest basins in the province, having developed only in the past 5–6 million years. Prior to that time, it was an area of low rolling hills, the eroded eastern portion of the ancestral Sierra Nevada.

Owens Valley is a down-dropped block bounded on both sides by normal faults. On the west is the Sierra Nevada Fault zone with 15,000–20,000 feet of vertical offset in the past 6 million years; on the east side, the White Mountain Fault zone has up to 26,000 feet of vertical offset over the past 12 million years. Running through the center of the valley is the Owens Valley Fault, along which the Sierra Nevada (west) side moves up and to the north relative to the rest of the valley. The basin is tectonically active, with the last major earthquake occurring on the Owens Valley Fault, sometimes called the Lone Pine Fault, in 1872. The magnitude of that earthquake is estimated at

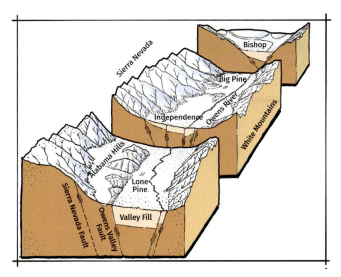

Block diagram showing the faults that underpin the Owens Valley (Adapted from Hollett et al., 1991, p. B21)

around 7.4 to 7.9; it created a scarp with up to 20 feet of vertical offset. I want to see this fault scarp where it is well exposed west of Lone Pine.

The town of Big Pine lies at the north end of a line of volcanoes and flows that erupted from another north–south fault. The Big Pine Volcanic Field covers about 190 square miles in the central Owens Valley. The field has over forty vents and cinder cones that erupted basaltic flows mostly 100,000–500,000 years ago. The most recent eruption, however, is thought to have been only about 17,000 years ago.

Like all the valleys in the Great Basin, the Owens Valley is internally drained, and it accumulates sediment eroded from all sides. The surface is mainly alluvial fan material. We can tell from water wells and seismic surveys that the basin fill near Lone Pine is up to 6,900 feet thick and consists of interfingering layers of alluvial fans, **talus**, volcanic flows and ash, river channels, glacial outwash, and lakebeds.

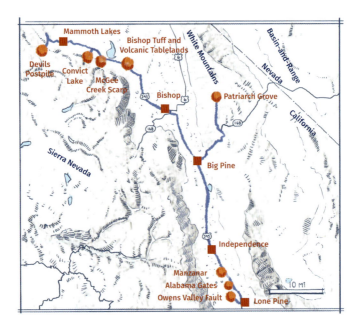

Map of the route from Devils Postpile to Lone Pine

In 1834, the Walker Party was the first group of Americans to enter the valley. This was still Spanish territory, before the Oregon Trail had been blazed. Andrew Jackson was president, and the westernmost state of the United States was Missouri. John Frémont, exploring the Louisiana Territory with a brigade of the US Army, traversed the Owens Valley in 1845. Frémont named it after Richard Owens, one of his guides. American settlers soon began moving into the valley and establishing cattle ranches. After the harsh winter of 1861–1862, the local game was devastated, and cattle began eating the crops the local Paiutes depended on. Starving, they began killing cattle, leading to the Owens Valley Indian War. By 1863, the settlers had forced the Paiute to move to Fort Tejon in the Central Valley south of Bakersfield.

As I continue driving south, there is more evidence of glaciation in the Sierra foothills to the west.

To get from Devils Postpile to Convict Lake, return on CA-203 to US-395 and drive south. The road to Convict Lake is 4.5 miles south of the intersection of CA-203/Mammoth Lakes Road with US-395 on the south side of the highway. Turn right (south) and drive to the Convict Lake parking area (37.595717, −118.850143), for a total of 22.8 miles (42 min).

STOP 39: CONVICT LAKE

Convict Lake, on the southern margin of the Long Valley Caldera, is dammed by terminal and lateral moraines. I pull over for a moment and see that the road goes through a low spot on the **terminal moraine**, the moraine formed at the farthest extent of a glacier. The abundance of boulders on the till surface suggests that the moraine is young, probably of Tioga age (about 15,000–27,000 years ago). East of Convict Creek, which flows out of the lake, is a 1,000-foot-high lateral moraine of Tahoe age (50,000–200,000 years old). The glacier that piled up the moraine had to be at least that thick.

Looking southwest from the lake, I can see brown **lenses** and blobs distributed randomly in the steep white cliffs. The bright cliffs are Mount Morrison Sandstone, whereas the rusty-brown rocks are Aspen Meadow Formation, a hornfels. The darker-brown blobs are Squares Tunnel Formation **argillite**, a hard, slaty rock. These are truly ancient rocks, Ordovician or Silurian in age, about 400–500 million years old, as best we can tell. Radiolarians and other rare fossils indicate that these were originally marine sands and muds. They are part of the Mount Morrison roof pendant, named for the peak just south of Convict Lake. These rocks have suffered a lot, first being squeezed and folded during the ancient Antler Orogeny, then being metamorphosed by heating as the Sierra Nevada granites engulfed them from below.

There is a fascinating backstory concerning how Convict Lake got its name.[9] In 1871, the east side of the Sierra was a pretty wild and lawless place. It was the frontier, and like most

View south along the road to Convict Lake. The road in this photo goes through a low spot in the terminal moraine (gray-brown) that dams the lake.

such places that did not yet have sufficient law enforcement, it attracted both pioneers (ranchers and miners) and outlaws (rustlers, gangs of former Confederate and Union soldiers). On a quiet Sunday, September 17, at the Nevada State Penitentiary in Carson City, twenty-nine prisoners overpowered their guards. They stole weapons and horses, and escaped. Contemporary accounts said they included murderers, rapists, train and stage robbers, cattle rustlers, and horse thieves.

Once out, the desperados split up. A group of thirteen headed south, robbing several locals and stealing more weapons, horses, and provisions along the way. Six of this group split off and continued south to escape the posse they were sure was following them. Their leader, a twenty-two-year-old murderer named Jones, had lived and worked in Mono and Inyo Counties and was familiar with the territory. Another convict, Roberts, was an eighteen-year-old stage robber who had grown up in Long Valley. Because Jones and Roberts knew the area, they planned to cross the Sierra to the western slope, where they felt they would be safe.

Unaware that the original posse had given up after two days, on September 19 they spotted a rider approaching. Fearing it was an outlier of the posse, Jones and Roberts ambushed him. The unlucky sod turned out to be a young Pony Express rider aptly named Billy Poor on his first mail delivery. Once they had him, they couldn't just let him go, and they couldn't hold him hostage because he would slow them down. So Jones shot poor Billy Poor just north of Bridgeport (20 miles north of Lee Vining).

When it became known that the convicts had murdered a local man, the good citizens of Mono County became enraged. By Friday, September 22, a new posse of ten men, led by Sheriff Robert Morrison and guided by an Indigenous deputy identified in the accounts as Mono Jim, was in hot pursuit. By Friday evening the posse caught up with the convicts camped by what was then called Monte Diablo Creek (today's Convict Creek). The posse went up Monte Diablo Creek toward where Convict Lake now lies. But the astute outlaws were in hiding, waiting on the south side of the creek. Sheriff Morrison was shot in the back of the head, killing him instantly. The posse retreated. Later the convicts came upon Mono Jim, who mistook them for the posse and called out to them. They ended up shooting him in the face. Jones and two others then fled south toward Bishop. Jones, with his knowledge of the countryside, disappeared and was never heard from again. The other two were eventually captured east of the White Mountains and returned to prison in Carson City.

A new posse was formed to track down Roberts and the others for killing Sheriff Morrison and Mono Jim. They finally captured the convicts just north of Bishop on September 27. On October 1 the prisoners were loaded on a wagon headed back to the jail in Carson City. It didn't get far before the wagon was surrounded by a large group of vigilantes. Without resistance, the guards turned over the prisoners, who were taken to a nearby cabin for a hasty trial. After a brief deliberation, the jury determined that two of the convicts should be hung at once and that Roberts, the young stage robber, would be sent back to jail. By November 15, eighteen of the twenty-nine convicts had been captured. Jones and ten others were never caught.

From that time on, the creek where Sherriff Morrison was shot became known as Convict Creek, and the lake feeding the creek has been called Convict Lake.

At McGee Creek I get a rare opportunity to see the mountain-front fault where it has displaced the glacial material and forms a scarp.

To get from Convict Lake to the McGee Creek scarp, return to US-395 and turn right (south). Drive to the McGee Creek turnoff and turn right (south); drive to the McGee Creek Campground parking area (37.563144, −118.785228), for a total of 8.3 miles (13 min).

STOP 40: MCGEE CREEK SCARP

Although this fault scarp is best viewed from above, there is visible offset that can be seen from the ground on the northwest side of the valley. On the southeast side, a line of green vegetation running up the moraine indicates groundwater coming to the surface along the trace of the fault.

I have not often seen a scarp developed in a glacial moraine, but that is exactly the case here. This does not tell me the total offset on the fault, but it does reveal that there has been 50 feet of displacement, up on the southwest side, *just since the moraine was deposited 12,000 years ago.* The scarp was caused by recent movement on the Hilton Creek Fault, a strand of the eastern Sierra Nevada Fault system. The Hilton Creek Fault had much more movement than suggested by the relatively small offset seen here. In places the scarp is about 3,600 feet high.

The base of the fault scarp is by a grove of aspen trees at the south end of the campground. The fault plane is perfectly aligned with the range front, demonstrating that this is part

The Hilton Creek Fault scarp is by a grove of aspen trees at the south end of the campground. McGee Creek Road also crosses the scarp about 650 feet from the campground turnoff. Note the cars in the parking lot, lower left, for scale. (US Geological Survey photo)

of the range-front fault system. The clean, planar fault scarp surface cutting the lateral moraine doesn't tell you how old the fault movement is, but is an indication of how recent the last fault movement was, and that the movement was rapid over a relatively short time.

To get from McGee Creek scarp to the Bishop Tuff, return to US-395 and drive south. Near Sherwin Summit there is a large pullout on the right (west) side of the highway (37.557693, −118.657231), for a total of 9.6 miles (13 min).

STOP 41: BISHOP TUFF AND VOLCANIC TABLELANDS, SHERWIN SUMMIT

I have it on good authority that this roadcut is visited by hundreds of geologists every year. They come to see the Bishop Tuff.

I pick up a piece of the pink tuff and notice how light it is. This volcanic ash is full of air that has been trapped between the ash fragments. This is a **welded tuff**, volcanic ash that was still hot when it fell out of the sky, and the hot fragments welded together. It is similar to pumice, another volcanic rock that is full of holes. The holes in pumice are formed from bubbles full of volcanic gases that came out of the lava as it rose to the surface.

The Bishop Tuff in this roadcut is a result of the collapse of the Long Valley Caldera. The accompanying eruption covered the entire region with the pinkish-tan welded tuff. The Bishop Tuff is the main reason for the Volcanic Tablelands north of Bishop. The broad, flat tablelands cover 325 square miles, and the tuff here is 490–660 feet thick. This is the main landform in the northern Owens Valley.

The lower part of the Bishop Tuff is a thick pumice layer. The upper section is welded ash. The welded layer is an **ignimbrite**, the result of a very hot cloud of ash and gas. This type of explosive eruption and incandescent mixture of ash and gas is called a ***nuée ardente***, the French term for

Bishop Tuff and sandstone dikes, Sherwin Summit, US-395

"flaming cloud." Because the mineral grains were welded together to form a hard and brittle layer, it is extremely resistant to erosion. That is why the tablelands, for the most part, are relatively flat and continuous.

I notice resistant ridges cutting diagonally across the tuff in the roadcut. It looks as though extension cracks, perhaps due to cooling and contraction, opened in the tuff. These openings allowed debris, mostly sand, to fall in. Later, with the help of mineral-carrying groundwater, the sandy debris became weakly cemented.

Local movements along faults caused minor tilting of the tuff. Because the welded layers are rather stiff, the tuff resists folding but is prone to fracturing. The mostly north–south faults are related to Basin-and-Range extension; they appear throughout the Volcanic Tablelands and appear to control stream erosion. The small hills scattered randomly across the tablelands are the result of ancient hot springs and steam vents that formed as the ash cooled and hardened.

At the south edge of the tablelands is the town of Bishop, a good place to tank up on gas and grab some lunch. Bishop, formerly Bishop Creek, is the largest city in Inyo County and the Owens Valley, with a population of 3,819 in 2020. Today the local economy is based on agriculture, mining, tourism, and cinema.

In the 1860s, the mining towns of Bodie and Aurora, just across the border in Nevada, were booming and needed beef to feed the miners. The nearest source of cattle was three hundred miles away in the San Joaquin Valley. The cattle would have to cross the southern Sierra at Walker Pass and move up the Owens Valley to the mines. Some of the cowboys noticed that the northern Owens Valley was ideal for raising livestock, and, to avoid the long cattle drive from the Central Valley, a few decided to settle in the valley. In 1861, Samuel Bishop, his wife, and several hired hands arrived in the Owens Valley with six hundred head of cattle and fifty horses. Bishop was one of the first European settlers in the valley. He built a homestead, the San Francis Ranch, along Bishop Creek and set up a market to sell beef to the miners in Aurora. By 1862, the town of Bishop Creek was established a couple of miles east of the San Francis Ranch.

I cast my eyes east from Bishop and see a largely treeless mountain range rising to 14,246 feet at White Mountain Peak, the highest peak in the Great Basin. These are the White Mountains, and they contain both the oldest living trees in the world and the greatest expanse of alpine tundra in the western United States.

To get to Patriarch Grove in the White Mountains, continue driving south on US-395 to Big Pine; at the northern outskirts of Big Pine, turn left onto CA-168 and drive 12.4 miles northeast to White Mountain Road on the left; turn left onto the gravel road and drive 22 miles farther to Patriarch Grove on the right (37.52717, −118.19743), for a total of 72.5 miles (2 hr 5 min).

If you would rather skip the Patriarch Grove stop and go directly from Bishop Tuff at Sherwin Summit to the Owens Valley

*Fault, continue south on US-395 to Lone Pine; turn right (west) onto Whitney Portal Road and drive 0.*7 miles west; turn right (north) on a gravel road by the No Water at Whitney Portal sign. Drive 0.6 miles north to the Owens Valley Fault (36.608645, −118.076915), for a total of 80.6 miles (1 hr 24 min).

STOP 42: BRISTLECONE PINES IN THE PATRIARCH GROVE, WHITE MOUNTAINS

I used to think that the giant sequoia, the most massive living tree, was also the oldest. Although a sequoia can be thousands of years old, it turns out that the oldest living tree, and perhaps the oldest living thing, is the humble bristlecone pine of the White Mountains. Imagine trees that were already 3,000 years old when the Roman Empire was at its peak. There are trees here that were well into their prime when the pyramids were being built, when Babylon was at its height, long before the Great Wall of China was conceived. Yet these are not mighty trees. They are not tall, and they are not massive. G*narly* is a more apt description. These weather-beaten sentinels have stood the tests of time, of a drying climate, of nutrient-poor rocky soil. These ancient trees have character, their twisted limbs and woody grain exposed to the weathering wind, searing sun, and extreme cold. It is well worth the drive, if you have the time, to one of the most secluded corners of California to see this primeval forest in all its unassuming glory.

As we approach the White Mountains from Big Pine, it is desolate and dry, very dry. No trees or shrubs of any kind. No communities, no structures. I am traveling with my son and daughter-in-law, and I am gaining new insights as we drive. My son's wife is a New York City girl, and she is getting anxious. "Where is everybody? Why is it so remote?" It is late in the year, and we are going to camp in one of the more isolated campgrounds. "What do you *do* when you go camping?" We try to reassure her that we are just going to put up a tent and walk around admiring old trees. "Please,

Bristlecone pines, Patriarch Grove, White Mountains

can we camp next to people?" There are three other people in the campground.

The sky is clear, and the sun is down by 6:30. Without the thermal blanket that clouds provide, it gets cold. At 8,600 feet, nighttime temperatures are in the low 30s. We have good sleeping bags, so we are comfortable. The great thing about camping in a place when there is no chance of rain is that you can throw your bag on the ground and sleep under the stars. This is considered a dark sky area, and boy is it. There are no lights anywhere for miles around, and the moon won't rise till much later. Lonely coyotes yelp and howl in the distance, and we can barely make out the dark forms of bats jinking after insects in the darkening sky. One by one, the stars come out to play. Eventually the Milky Way is splashed from one horizon to the other, an arching, glowing cloud. And shooting stars by the dozens etch their tracks across the sky.

A few words about the rocks are warranted. The White Mountains are an uplifted fault-bounded range typical of the north–south elongated mountains in the Basin-and-

Range. The basement rock is granite, roughly the same age as the Sierra Nevada granite. The granite intruded into Precambrian–Cambrian-age marine shale, sandstone, and **dolomite**. It is the light-colored dolomite that gives the range its name. Above the granites are Miocene volcanic rocks that were tilted to the east when the White Mountains were uplifted around 12 million years ago. After uplift there was another outpouring of Miocene and Pliocene volcanic rocks. The bristlecones are growing primarily on white dolomite, a hard carbonate rock similar to limestone that formed in shallow tropical seas over 500 million years ago. It is a tough, rocky soil. And arid—visitors are warned to bring their own water, as there is none available.

We leave behind these timberline ancients. I deposit my son and his wife in Big Pine, then continue driving south. The remoteness of the Owens Valley between Bishop and Lone Pine lent itself to a sad chapter in the history of our country, the unjust incarceration of Japanese Americans during World War II. Manzanar, five miles south of the town of Independence on Highway 395, was the site of a World War II Japanese American prison camp. (In America we don't like to use the term *concentration camp*, although many have called it just that.) It is a now a National Historic Site that commemorates the injustice of imprisoning American citizens only because of their ethnic background.

In early 1942, shortly after the attack on Pearl Harbor, the US and California governments, driven by fear and greed, ordered more than 110,000 men, women, and children to leave their West Coast homes and move to remote camps surrounded by barbed wire and guard towers. Camp residents lost an aggregate $400 million in property during their incarceration: it was either stolen or sold off. The Manzanar War Relocation Center was one of ten camps where Japanese American citizens and resident Japanese aliens were interned during World War II. Between 1942 and 1945, Camp Manzanar in the Owens Valley was home to as many as

Manzanar prison camp at dusk (Photo courtesy of Alexander Novati, Wiki Commons, https://commons.wikimedia .org/wiki/File:Manzanar_Internment_Camp.jpg)

ten thousand Japanese Americans. Their stories have been told in numerous books, most famously in the 1973 memoir *Farewell to Manzanar* by Jeanne Wakatsuki Houston and James D. Houston. The federal government did not apologize for the internment until 1988. California didn't apologize until 2020.

Continuing south, I come upon a memorial to the struggle for water in this parched land. Five miles south of the Manzanar turnoff, on a nondescript hillside west of US-395, and marked by a small building with an American flag, is Alabama Gates. This is a site made famous by the Los Angeles–Owens Valley "Water Wars."[10]

Los Angeles, ever expanding at the turn of the twentieth century, needed water for its orange groves and never-ending influx of new residents. Led by William Mulholland, a self-taught civil engineer, the city began buying up water rights across California, including in the Owens Valley. In the early 1900s, a large aqueduct was built to bring Owens Valley water south to Los Angeles. The aqueduct began moving water from the Owens River to LA in 1913.

Construction of the LA Aqueduct, 1912
(Unknown photographer, Wiki Commons)

During the 1920s, there were several years of lower-than-normal snowfall in the eastern Sierra, and local water needs were increasing. By the spring of 1923, both Los Angeles and the Owens Valley were facing water shortages. In order to increase supply, the LA Department of Water and Power (LA DWP) began pumping groundwater in the Owens Valley. Farmers around Independence filed injunctions to halt the lowering of the water table. Residents of Bishop and Lone Pine became alarmed by LA's purchases of groundwater rights north of Independence.

Wilfred and Mark Watterson, owners of the Inyo County Bank, organized valley residents to form an irrigation district to oppose the purchase of water rights by Los Angeles. This led to a series of escalating confrontations. Farmers illegally diverted water, leaving the canal empty. The LA DWP bought up land and water rights indiscriminately. Area farmers felt vulnerable, with some neighbors selling out to LA and others maintaining their water rights.

Violence erupted in May 1924, when forty men dynamited the Lone Pine Aqueduct spillway gate. No arrests were ever made. In November 1924, Mark Watterson led about sixty to one hundred people to occupy the Alabama Gates, the gatehouse that diverted Owens River water into the aqueduct. They closed the aqueduct by opening an emergency spillway. The LA DWP began negotiations to buy farms and provide compensation, which ended the occupation. Attacks on the aqueduct began again in April 1926, and by July 1927 there had been ten cases of dynamiting the aqueduct. Then the Wattersons suddenly closed all branches of the Inyo County Bank: they not only were bankrupt but were convicted on thirty-six counts of embezzlement. In the face of the collapse of both resistance and the Owens Valley economy, LA sponsored a series of aqueduct repair and maintenance programs that stimulated local employment. That was the end of it. For a while.

In 1983, a teenager, inspired by the Owens Valley water wars, damaged the Alabama Gates again. A judge sentenced him to attend college.

The legacy of the water wars is that there is virtually no agriculture remaining in this valley. It is classic Mojave Desert, with sunbaked hills and barren rock outcrops separated by wide swaths of sage-covered flats. It is easy to see why, along with the snowy peaks of the Sierra, the area makes a great backdrop for Western movies.

Six miles south of Alabama Gates is Lone Pine, renowned for its spectacular setting at the base of the High Sierra, for the Alabama Hills that have been featured in so many classic Westerns, and for the Museum of Western Film History. Located at an elevation of 3,727 feet, this scenic outpost had a population of 1,484 in 2020. The town was named after a solitary pine at the mouth of Lone Pine Creek, which drains the Sierra beneath Mount Whitney.

The Lone Pine area is one of several originally inhabited by the Timbisha band of the Shoshone tribe, also known as the Panamint or Koso. The tribe is now based at Furnace Creek in Death Valley.[11]

A cabin was built here during the winter of 1861–62, and a settlement developed over the following two years. A post office opened in 1870, making the town official. During the 1870s, Lone Pine was an important supply stop for nearby mining communities, such as Cerro Gordo. At one time the Cerro Gordo Mine, high in the Inyo Mountains to the east, was one of the richest silver mines in California.

In 1920, the fate of Lone Pine changed forever when a movie production company came to the Alabama Hills to make a silent film, *The Round-Up*. Other companies soon discovered the scenic location, a jumble of bald granite knobs poking out of the desert. Over the next several decades more than four hundred films, one hundred television episodes, and countless commercials were filmed here. Today tourism is a mainstay of the economy, because Lone Pine is the southern gateway to the eastern Sierra.

Along the east side of the Alabama Hills, just west of Lone Pine, is my objective, the Owens Valley Fault scarp.

To get from Patriarch Grove to the Owens Valley Fault, return south on White Mountain Road to CA-168; turn right (west) on CA-168 and drive to US-395; turn left (south) on US-395 to Lone Pine; turn right (west) onto Whitney Portal Road and drive 0.7 miles west; turn right (north) on a gravel road by the No Water at Whitney Portal sign. Drive 0.6 miles north to the Owens Valley Fault (36.608645, –118.076915), for a total of 77.9 miles (2 hr 11 min).

STOP 43: OWENS VALLEY FAULT

The trace of the Owens Valley Fault is still evident in the outwash gravels west of Lone Pine. A 10–13-foot-high north–south-trending escarpment displaces the valley surface. As with the Sierra mountain-front fault, this much smaller scarp is up on the west side and down on the valley side.

The break in slope here is the most visible sign of the largest-ever recorded earthquake to hit this area, which took place in 1872. Another reminder of the great shake is the cemetery for victims of the quake located about a mile north of downtown Lone Pine. The cemetery is located just west of US-395 on the main fault scarp.

Seismologists have shown that longer scarps mean that there were bigger earthquakes. And this was a big one. Centered on the Owens Valley Fault, the temblor had an estimated magnitude of 7.4–7.9, based on the reported damage and the distance over which the quake was felt. This is the same magnitude as the 1906 San Francisco earthquake, although there were far fewer people living in the Owens Valley. Most of the buildings were made of adobe mud

The Owens Valley Fault scarp (between the arrows) west of Lone Pine. View north.

bricks, and sixty of the eighty structures in the area crumbled. Twenty-six people died, most in collapsed buildings, and they were buried in a mass grave at the north end of town. The resulting scarp at the time was described as 15–20 feet high, although that may have been the cumulative offset of three or more quake events. The break is nearly 60 miles long, extending north from Owens Lake to near Big Pine.

The Owens Valley Fault is part of a system of faults that dropped the Owens Valley to its current elevation. It is also partly responsible for uplifting the Alabama Hills, the part of the scenery that makes me feel as though I just drove onto a Western movie set. Which is, in fact, what I have done.

To get from the Owens Valley Fault to Movie Flats, Alabama Hills, return south to Whitney Portal Road and turn right (west); drive an additional 2.0 miles (3 min) west to Movie Flat Road; turn right (north) and drive 1.6 miles to Mobius Arch Loop Trailhead parking area on the left (36.611193, –118.124545).

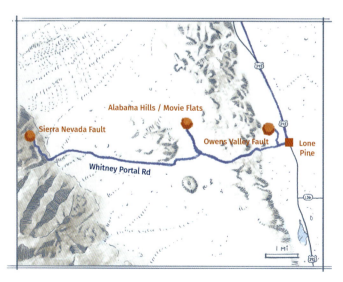

Map of the route from Lone Pine to the Alabama Hills and the Sierra Nevada Fault

STOP 44: MOVIE FLATS, ALABAMA HILLS

Despite being a continent away from the main battles of the Civil War, the Owens Valley had its share of Union and Confederate sympathizers. The Alabama Hills were named by rebel supporters after the CSS *Alabama*. When the *Alabama* was sunk by the Northern USS *Kearsarge* in 1864, prospectors sympathetic to the North named a mining district, a mountain pass, a peak, and a town after the *Kearsarge*.

The Alabama Hills are bounded by the Owens Valley Fault on the east and the Sierra Nevada Fault on the west. The hills form the western backdrop of Lone Pine. They are a highly fractured north–south-elongated uplifted block of 85-million-year-old (Cretaceous) light-gray Sierra Nevada granodiorite. There are also patches of red-brown Triassic–Jurassic metavolcanics and metasediments. The faults pushed up the hills over the past 4–6 million years. Over the past 2 million years, the hills were blanketed in glacial outwash from at least three glacial periods.

Mount Whitney (center right) as seen from the Alabama Hills. That it does not appear to be the highest peak on the ridge is a quirk of perspective. This scenic area has served as the backdrop for many Hollywood westerns. (National Park Service photo)

Movie Flats in the Alabama Hills is one of the classic movie settings of all time, right up there with Monument Valley and the Grand Tetons. The scenic fractured-granite hills served as the background in unforgettable television shows I watched as a child (and here I risk dating myself), including *Hopalong Cassidy*, *The Gene Autry Show*, *Wagon Train*, *Bonanza*, *Sky King*, *Have Gun Will Travel*, *Death Valley Days*, *Annie Oakley*, and *The Lone Ranger*.

Movies have been filmed here for over a century. The first to be filmed in the Alabama Hills are *The Roundup*; *Water, Water Everywhere*; and *Cupid, the Cowpuncher*, shot in 1919 and 1920. More memorable movies include, among many others, *Riders of the Purple Sage*, *The Lives of a Bengal Lancer*, *Westward Ho*, *Charge of the Light Brigade*, *The Cisco Kid and the Lady*, *Gunga Din*, *High Sierra*, *Heart of the Golden West*, *The Oxbow Incident*, *The Gunfighter*, *Broken Arrow*, *Rawhide*, *Springfield Rifle*, *Destry*, *Bad Day at Black Rock*, *The Law and Jake Wade*, *How the West Was Won*, *The Great Race*, *Zabriskie Point*, *Joe Kidd*, *Farewell to Manzanar*, *Tremors*, *Gladiator*, and *Django Unchained*.

The Alabama Hills are mere stepping stones compared to the mountains raised by the fault at my next stop.

To get from Movie Flats to the Sierra Nevada Fault, return south to Whitney Portal Road and turn right (west); drive a total of 7.9 miles (15 min) and pull over at the bend of the first switchback (36.605186, −118.215822).

STOP 45: SIERRA NEVADA FAULT

There was an excellent view of the ramparts of the eastern Sierra Nevada Fault scarp from the turnoff to Mammoth Lakes. Here I am just below Mount Whitney, the culmination of the range, the highest part of the High Sierra. At this stop, I cross the actual fault and climb about five hundred feet up the range to get a better view of the massive rent in the Earth that pushed up the mountains. I am just below the tree zone, with a few scraggly pines scattered on the steep slopes. Below me the sparse, scrub-covered plains of the valley bottom stretch north and south into the distant haze. The actual trace of the fault is at the foot of the slope, just before the road starts climbing, but this stop provides a great view north and south along the escarpment of the Sierra Nevada Fault. The evidence of the scarp is clear: west of the trace, there is granite at the surface; east of the scarp is dirt, alluvial valley fill. West of the scarp, streams cut deep valleys; on the east side, the streams meander across a gentle plain. And every so often along the scarp there is a cluster of dense vegetation—cottonwoods and willows that have tapped into the groundwater bubbling up near the surface along the fault. Groundwater has a hard time moving through massive granite, but it has no problem moving in the zone of broken rock encountered along faults.

There is 10,000 feet of elevation change between the Owens Valley bottom and the top of Mount Whitney. And yet that is just part of the total that the boundary fault has uplifted the granite peaks. Much of the displacement along this fault is below the surface. Another 4,000 feet of uplift are buried beneath the valley fill sediments. The balance of the roughly 20,000 feet of uplift has been eroded from the top of the Sierra. No matter how you cut it, that makes this one of the great fault escarpments of the world.

Whitney Portal, a few miles up this road, is the jumping-off point at the south end of the John Muir Trail. It is also the east end of the High Sierra Trail that crosses the Sierra from west to east. I hiked this trail when I was seventeen, and climbed Whitney the easy way, from the west. Those intrepid hikers that choose to climb Mount Whitney (14,505 feet) from Whitney Portal can thank the Sierra Nevada Fault for every inch of the 6,600 feet they have to climb.

Scarps belonging to various strands of the Sierra frontal fault zone can be seen slicing across alluvial fans from here north to Independence. The main fault segment in this area

Basin-and-Range   211

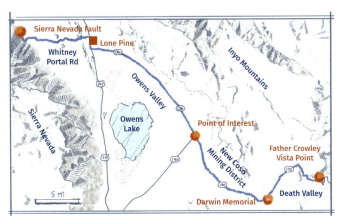

Map of the route from the Sierra Nevada Fault and Lone Pine to Father Crowley Vista Point, Death Valley National Park

is called the Independence Fault. There has been no movement on the Independence strand of the Sierra Nevada Fault in historical times.

This fault zone marks not only the boundary between the Sierra and the Owens Valley. As we saw at Mono Lake, it also separates the Sierra Nevada and the actively extending Basin-and-Range province. The eastern margin of the Basin-and-Range province, marked by the Wasatch Front at Salt Lake City, lies 400 miles to the east.

To get from the Sierra Nevada Fault to Father Crowley Vista Point, return to US-395 at Lone Pine and turn right (south); turn left (east) onto CA-136 and drive to CA-190; continue straight on CA-190 to Father Crowley Vista Point on the left (36.351898, −117.550430), for a total of 51.3 miles (57 min).

From Lone Pine in the Owens Valley, I head south and east toward Death Valley. After a few miles I'm driving along the eastern shore of Owens Lake. Except that it's not really a shoreline, because it's not really a lake anymore. It's a playa, a dry lakebed that holds water only after a good rain or an exceptional snowmelt.

A large freshwater lake filled much of southern Owens Valley during the last glacial episode. The highest **lakeshore terraces** indicate that between about 27,000 and 15,000 years ago, glacial Owens Lake was as much as 300 feet deep and extended north to Independence. Like Mono Lake, Owens Lake has been shrinking since the end of the last ice age. More recent shorelines near the north end of the lake indicate a water depth of 90 feet. The drop in lake level is a result of the arid climate that developed here after the last ice age, and because the peaks of the Sierra Nevada capture most of the moist air heading east from the Pacific and wring all the water out of it as snow and summer thunderstorms. But the story doesn't end there. The drop is not only because of the climate and the rain-shadow effect. Once again, it has to do with Southern California's unending quest for water.

When Euro-Americans started settling this part of the Owens Valley in the 1880s, the lake was about 30 feet deep. At that point the water level began dropping because farmers were capturing water from the streams feeding the lake and running it through irrigation ditches to water their crops. In 1900, the local farmers were using 500,000 acre-feet of water every year. (An acre-foot is the amount of water it takes to cover one acre with one foot of water.) Then, in 1913, the 223-mile-long Los Angeles Aqueduct was completed. It began moving 350,000 acre-feet of water per year from the Owens River to Los Angeles. The Owens River was the main inflow to Owens Lake. The combination of local irrigation, the aqueduct, and evaporation was more than the inflow to the lake, and by 1926 the lake had dried up completely.

Water rights had been disputed for many years, but eventually the LA DWP controlled all of the local water rights and almost completely diverted the Owens River to Los Angeles. For those who remember the 1974 movie *Chinatown*, the plot was inspired by LA's voracious need for water, the building of the LA Aqueduct, and the Owens Valley water wars. As Noah Cross, the wealthy industrialist

Historical photo of Owens Lake while it still had water (Photo courtesy of Rich McCutchan)

and crime boss in the movie said, "Either you bring the water to LA or you bring LA to the water." They brought the water to LA. Not content with the surface water, the LA DWP completed a second aqueduct in 1970 to divert much of the valley's groundwater south to Los Angeles.

As a result of many lawsuits, an understanding was eventually reached between the Owens Valley Committee (and Inyo County, the Sierra Club, and others) and the LA DWP. Los Angeles agreed to restore parts of the Owens River. It took a while, but by 2008 this was being implemented. Some water occasionally makes it into the lake. For example, in 2023, the lake flooded for the first time in over a hundred years due to the numerous storms that struck California during the first three months of the year, increasing the lake volume to about 50,000 acre-feet.

Apparently, a lake has been in this location for quite some time. Wells drilled into the dry lakebed indicate that there are up to 7,000 feet of lake sediments, mainly clay and silt, above the granite basement rock.

Driving by the northeast shore of the former lake, I see solar evaporation ponds. (Solar evaporation ponds are shallow artificial basins that use sunlight to evaporate water and produce salt or other minerals.) Some of the flow from the Owens River has been restored, and the lake now contains a little water that is used to extract salts from the lakebed. Pink areas are salt-encrusted waters colored by halobacteria, a type of microbe that requires high concentrations of salt to survive. Dark areas are green-colored mud, inhabited by a salt-tolerant green algae. The white deposits are trona, a mineral that is refined into soda ash, which is used to make glass, paper products, laundry detergents, and other products such as sodium bicarbonate (baking soda) and sodium phosphates (detergents). In 1962, Morrison and Weatherly Chemical Corporation began mining trona from the dry lakebed. Small-scale mining continues today.

Despite there being at least a small amount of water in the lake, as of 2013 Owens Lake was the largest single source of dust pollution in the United States. Environmental and Indigenous groups are working to keep more water in the lake to reduce the dust.

Driving south and east from Lone Pine, I cross the south end of the Inyo Mountains. In this area, the rocks are mainly Pennsylvanian and Permian sedimentary and metasedimentary rocks (limestone, mudstone, fine-grained sandstone, and hornfels) invaded by Jurassic igneous rocks. The whole sequence was then covered by a blanket of resistant 4–7-million-year-old (Tertiary) basalts that now form the flat-topped Malpais Mesa.

At the junction of CA-136 and CA-190 is a sign for a "Point of Interest" (36.428947, -117.825265). I pull over briefly here just to see the grand sweeping panorama of the Sierra Nevada to the west. An interpretive display names the peaks along the Sierra Crest, gives a brief history of roads and mining in the area, and explains the role of tourism in the local economy.

Thirteen miles beyond the Point of Interest is the Darwin Historic Marker on the right (south) side of the highway (36.318241, −117.671405). I'm thrilled. Could it be that Darwin was actually in California? I don't recall ever hearing anything about that.

That's because it has nothing to do with Charles Darwin, who was never anywhere near here. Rather, it commemorates the town of Darwin five miles off the main road south of here in the Coso Range. The town was named after Dr. Darwin French, who explored the area in 1860 and lent his name to several landmarks. The settlement was the center of activity for the New Coso Mining District, a cluster of mines that produced mostly silver but also some gold, lead, zinc, copper, and mercury. The first mine, the Promontorio, was located in October 1874, allegedly by a prospector looking for a lost mule. By December of that year, two hundred men had flocked to the district and started digging holes. It didn't take long to find metal.

By 1875, the mines had produced some six thousand bullion bars worth $100,000, from mines less than 100 feet deep. Each day, the smelter furnaces would be fed 20 tons of ore, along with 1½ tons of iron ore, 4 tons of **slag**, and 3 tons of lime. If the ore was of high-enough grade, this 29-ton mixture produced 6 or 7 tons of bullion (150–175 bars) worth $2,000 in silver alone. By the end of 1875, Darwin had three smelters, twenty mines, and a Wells Fargo express office. All water and timber for the seven hundred residents had to be brought in. By 1876 the population reached one thousand. Most production came from the Christmas Gift, Defiance, and Lucky Jim Mines. By 1888 the district was mostly played out—all the good ore was depleted.

During this brief mining boom, copper had been left on the mine dumps because it was considered worthless. But around 1906, electric lighting was all the rage. Copper was in high demand for electrical wiring and had become valuable. This led to a second boom in the district, and by 1907, the Lucky Jim Mine was shipping copper and lead-silver ore to smelters in Salt Lake City.

By 1927, the Darwin Mining District was producing silver, lead, gold, tungsten, and copper. In 1945, the Anaconda Copper Company bought the main mines and took over operations. Darwin became the leading source of lead in California. Alas, the late 1940s were the high point for the district, and by 1952 the mines and town were largely abandoned. Total production from the district between 1875 and 1951 was about 7.6 million ounces of silver, over 58 thousand tons of lead, and over 700 tons of copper. Taken all together, metals worth $29 million were extracted from the ground here.

The ore bodies in the New Coso Mining District occur as fissure-filling and limestone-replacement-type deposits in the thick (up to 4,000-foot) Keeler Canyon Formation limestone of Pennsylvanian–Permian age. First the limestone was metamorphosed and altered at the contact with the Jurassic-age Darwin **quartz monzonite** (an igneous relative of granite). The limestone is very reactive to acid-rich hot fluids coming off the magma, so near the zone of contact the limestone was replaced by galena (lead and silver), sphalerite (zinc), chalcopyrite (copper), and other metallic minerals dissolved in and carried by the magmatic liquids. Later these rocks were cut by faults, and more metallic minerals precipitated as veins in open spaces (fissures) along the faults.

Today there's not much to show for the frantic activity of the mining district. The former mining camp is a ghost town with a few old wooden structures straining against gravity to stay upright, along with a handful of lived-in homes, an open-air art museum, and a post office set against a backdrop of dry grass, barrel cactus, Russian thistle (tumbleweeds), and the occasional Joshua tree.

And it's only getting hotter and drier the farther east I drive.

DEATH VALLEY NATIONAL PARK

Death Valley is one of the few places on Earth where the ground is literally tearing itself apart. As the land split along faults, the valley dropped down between the mountains. The mountains, in turn, bulged upward. As the mountains rose, on some of them the upper parts slid off into the adjacent valleys. But more about this later.

January through April is the best time to visit Death Valley, in my opinion. I'm here in February, and it is delightful: high temperatures are in the 60s and 70s under clear blue skies. After April it is a blast furnace. On July 10, 1913, Oscar Denton, a US Weather Bureau observer stationed at Furnace Creek, claimed the mercury hit a blistering 134°F. According to *The Guardian*, this is the highest temperature ever recorded on Earth. Furnace Creek regularly gets into the 120° range, an eye-stinging, scorching heat that has been described as being in an oven or having a blow-dryer in your face.

In early spring the weather is good. The wildflowers are at their peak from late March to early April. Temperatures are in the mid-60s in January, mid-70s in February, and low 80s in March. Death Valley averages only 2¼ inches of rain in an entire year, and most of that falls from January to March.

The unique climate, landscape, and geology of this region led President Hoover to declare Death Valley a national monument in 1933. The United Nations designated it a biosphere reserve in 1984. Believe it or not, there are animals that live here, from desert pupfish to desert bighorns and herds of feral burros left behind by prospectors. Congress upgraded it to a national park in 1994. In 2013, it was named a "dark sky park" by the International Dark Sky Association. I do not recommend visiting in the summer. And, as it is a desert, winters can be freezing cold, too. Whenever you go, don't wander too far from a road unless you know exactly what you are doing.

Long before Euro-Americans arrived, a group of Shoshone, the Timbisha band, roamed the area from Death Valley to Owens Valley, migrating from valleys to mountains with the seasons. Their name is derived from *tümpisa*, the red-ochre earth sourced by the Old Ones from the Black

Hills overlooking Furnace Creek. This iron oxide–based pigment was smudged on faces and used in homes for spiritual purposes. The people called their ancestral homeland Tüpippüh, and anthropologists say they have occupied the region for over a thousand years. The Timbisha say they have been here forever.[12]

In the winter of 1849–1850, a group that was part of the California Gold Rush headed west from Salt Lake City too late to make it over the Sierra before the first snow. So they headed southwest, looking for the Old Spanish Trail, a way around the south end of the mountains. They ended up stuck near Furnace Creek in December with little water, no forage, and no supplies, sure they would all die there. Although only one person did die, when they finally found their way out a month later, they reportedly said "Goodbye, Death Valley." The name stuck.

Death Valley is a fault-bounded basin that formed in the past 14–15 million years. But the rocks are much older. The oldest formations are 1.4–1.8-billion-year-old metamorphic rocks of uncertain origin. Following a long period during which the metamorphic rocks were eroded, marine sediments then accumulated in this area along the western margin of proto–North America. The area was near the equator, and tropical limestone deposits similar to the Bahama Banks formed, along with occasional sandy beaches and muddy bays. These sediments range in age from late Precambrian through Pennsylvanian (approximately 300–600 million years old).

During that time, these rocks were subjected to deformation at least twice. The first time was during the Antler Orogeny, which lasted from about 320 to about 370 million years ago. The Antler event involved uplift and eastward thrusting of great sheets of rock. The same thing happened again during the Sonoma Orogeny around 250 million years ago. Both of these periods of deformation are related to convergence and subduction of oceanic crust beneath continental crust that was happening west of here. This was long before the Franciscan subduction that formed most of western California.

During the Mesozoic, about 145–180 million years ago, the sedimentary layers in this area were invaded by regional magma bodies, similar to Sierra Nevada granites, and were again thrust eastward during the Sevier Orogeny, some rocks moving as much as 45 miles to the east on large regional faults. This period corresponds to early Franciscan subduction.

In early Cenozoic time (about 25–67 million years ago), the area was eroded almost to sea level. The sediments were washed to the sea; none remain from that period. Starting in late Cenozoic time, perhaps as far back as 15–20 million years ago, Death Valley underwent regional uplift and Basin-and-Range extension, faulting, and basaltic volcanism. Up-arching of deeply buried crustal rocks led to low-angle faulting where gigantic blocks of sedimentary rock slid off the metamorphic cores of the uplifted mountains. The peculiar domed shape of these **metamorphic core complexes** led to their being named **turtleback structures**.

We know that around 25 million years ago the East Pacific Rise, the mid-ocean spreading center, reached the subduction zone and began to be overridden by the continent. Now the continent, affected by the spreading center beneath it, started to extend. Rifted basins formed in the Salton Trough (from the Salton Sea to the Colorado River delta) and Death Valley areas around 15 million years ago. At the same time, strands of the San Andreas Fault system caused sideways movement along faults in the Death Valley area. The Furnace Creek and Sheephead Faults that bound the valley have both vertical and horizontal displacement. As the faults moved, they caused the area between them to drop. Stretching of the crust created pathways for magma to rise to the surface along the faults. Volcanic features found in the park include black basalt flows, volcanic ash deposits, and cinder cones. Explosion craters mark where magma encountered groundwater that flashed to steam: one example is Ubehebe Crater. The crust is still stretching here: the youngest volcanic activity occurred only 2,000 years ago.

During the last ice age, Death Valley was filled by glacial Lake Manly. Obviously, the climate was both cooler and wetter. The lake was deepest between 128,000 and 185,000 years ago; at that time, it covered 620 square miles to a depth of 580 feet. Death Valley, like the Owens Valley, is an internally drained basin. Here, the water—when there is any—comes mainly from the Amargosa River. After the early Lake Manly dried, another lake formed sometime between 10,000 and 35,000 years ago. It was only 30 feet deep. This lake, too, gradually dried up. Minerals dissolved in the lake water became more and more concentrated as the water evaporated. Only a salty brine or salt crust is left in the deepest parts of the valley, in places such as Badwater. Today, there is water in the valley only after strong rains. In February 2024, the National Park Service reported that heavy rains had reconstituted Lake Manley and that it was about 6 miles long, 3 miles wide, and 1 foot deep. By the end of April, the lake was gone again.

As with most areas in the West, the first Americans came here to make their fortune. Fur trappers didn't bother to come here, as there are no rivers. But prospectors and miners did come. Three commodities were mined in Death Valley: gold, **talc**, and **borax**. Believe it or not, the valley is famous for the borax. But we'll get to that.

Talc, the softest mineral, is known for its silky texture. It's so soft that a fingernail can scratch it. The silkiness and softness are due to its crystal structure, which enables the very fine layers to slip easily over each other. This property makes it useful in a variety of products, such as cosmetics and paints. As a powder, it absorbs moisture well and helps cut down on friction, which is why baby powders used talc to keep skin dry and help prevent rashes.

In its natural form, some (but not all) talc contains **tremolite**, a form of asbestos. As we all know, asbestos can cause cancer and other lung diseases when inhaled as a powder. Although talc is not banned in baby powder, many baby powders today use corn starch as an absorbent instead.

Talc deposits are found in the Precambrian Crystal Spring Formation, a dolomite. Dolomite is a type of limestone that contains magnesium: it forms when magnesium-rich groundwater encounters a limestone (calcium carbonate), and the magnesium replaces the calcium. The Crystal Spring dolomite was altered by heating when magma was injected between the rock layers. The altered dolomite contains talc deposits that outcrop along a 50-mile-long belt in the Panamint Mountains and the Kingston Range of southern Death Valley. The talc zone appears as a bright white stripe 30–100 feet thick; the talc consists of elongated lenses or pods that can be traced for 3,000 feet or more.

Talc was first mined in Death Valley in 1910; since the mines existed before the park, they were grandfathered into the national park and allowed to continue mining. The Park Service closed all the mines in 1980 due to the asbestos scare.

Gold was also mined in Death Valley. In the winter of 1903, Jack Keane and Domingo Etcharren were prospecting for silver in the area. They were local celebrities for having found nothing in eight years of searching. Then Keane stumbled on an outcrop of visible gold in the Funeral Mountains north of Beatty Junction. The resulting mine was named the Keane Wonder Mine because local wags thought it a wonder that anything had been found. It began a small gold rush. In 1906, Homer Wilson bought the mine and built a stamp mill and aerial tramway up the Funeral Range to the mine. Fifty men were working the mine and mill by 1909. The mine produced off and on until 1916, when it closed for good as the ore played out. The mine workings included a 246-foot-deep shaft and over 6,500 feet of tunnels. Most of these are not accessible to the public, and wouldn't be safe if they were.

The Wonder Mine is in the Chloride Cliff Mining District. The main rock type in the area is Precambrian Crystal Spring Formation, the same dolomite that the talc came from. Gold occurs in quartz veins that follow the surfaces of rock layers, especially along the contact between the metamorphosed dolomite and igneous layers. The veins extend at least 1,650 feet and are 6–26 feet thick. Pyrite (fool's gold) and galena (lead-silver ore) were found along with gold.

The Keane Wonder Mine produced a total of about 35,000 ounces of gold. In the 1970s, the mine was bought by the National Park Service and preserved as a historic landmark. It is one of the two largest gold mines in Death Valley, the other being the Skidoo Mine.

Gold was discovered in 1906 at Skidoo, on the west side of the Panamint Range near Emigrant Canyon, by two miners on their way to the nearby settlement of Harrisburg. The population quickly grew to around seven hundred, but all had left within ten years as the ore ran out. The Skidoo Mining District produced about 75,000 ounces of gold. Skidoo had a post office, newspaper, school, and bank, but the last buildings were gone by the 1980s. The townsite is now covered by the ubiquitous sagebrush, although fragments of glass and masonry can still be found. The surrounding hills have over a thousand mine entrances and workings, plus dozens of scattered structures—weathered wooden cabins, rusting **headframes**, broken-down iron machinery, and old vehicles. The stamp mill is still largely intact.

This brings us to borax. In the days before soap was common, borax and borate minerals were used as a detergent, cleaning agent, and antiseptic. In 1881, borax was discovered in the dry lakebed at the mouth of Furnace Creek Wash in Death Valley. The white crystalline mineral **ulexite**, the main borax ore, was known to the miners as "cottonball" because it occurred as cottonball-sized white nodules. Borax miners would just scrape the mineral off the

A twenty-mule team in Death Valley (National Park Service photo)

dry lakebed. The Eagle Borax Works began operating in 1881, followed by the Harmony Borax Works in 1882. Within a few years, these operations were producing 3 tons of borax a day. The ore was dissolved in boiling water, and pure borax would precipitate out as the solution cooled. The borax was shipped in gigantic wagons, 16 feet long by 6 feet wide and weighing upward of 36 tons fully loaded. It required teams of twenty mules to haul these massive wagons 165 miles across the desert to the railhead at Mojave. Between 1883 and 1888, over 12 million pounds of borax were hauled out of Death Valley. Borax was carried primarily by the twenty-mule teams until the Borate & Daggett Railroad was built into the valley around 1895. The Dial Corporation still markets 20 Mule Team Borax as a laundry detergent.

Everyone has heard of Death Valley, if for no other reason than the old television program *Death Valley Days*. Originally a radio program created in 1930, then a television series from 1952 to 1970, the show told true stories of the Old West centered around Death Valley. It is considered one of the longest-running Western programs in broadcast history. The series was sponsored by Pacific Coast Borax Company, makers of 20 Mule Team Borax detergent and Boraxo hand soap. It was hosted by Ronald Reagan (among others), who left the series to run for governor of California in 1966. Pacific Coast Borax merged with United States Potash to form U.S. Borax in 1956. Its borax-based product line was sold to Dial Corporation in 1988. U.S. Borax still produces borax from the Rio Tinto Borax Mine, the largest open-pit mine in California, 60 miles southwest of the national park, near the town of Boron, California.

I have never been to Death Valley, but have heard all kinds of superlatives: the hottest, driest, and lowest place in North America. I want to see what people got all excited about. The park entrance is 5 miles east of the Darwin Historic Marker on CA-190.

STOP 46: FATHER CROWLEY VISTA POINT

Situated a few miles inside the park boundary, the Father Crowley Vista Point sits on a high point that overlooks the northern Panamint Valley. The stop is named for the "Desert Padre," Father John Crowley, a Catholic priest who fought for local water rights and promoted tourism in the area during the 1920s and 1930s.

This is a multihued moonscape of barren volcanic rock and gravel. It may sound odd, but there is a stark kind of harsh, austere, baked beauty to the landscape. Of course, I like it because all the rocks are exposed. There are no trees, no shrubs, no grass; there is almost no soil to obscure or cover these bones of the Earth. I am standing on Miocene pyroclastic and volcanic mudflow deposits, the outpourings of nearby volcanoes. A glance into Rainbow Canyon shows what lies below.

Rainbow Canyon runs along the north side of the parking area. It is a steep, 1,000-foot-deep gash where black Pliocene (2.5–5-million-year-old) basalt caps the plateau. Identifying the rocks below the basalt is not as easy: perhaps because of the tough terrain, geologists have not spent a lot of time unraveling the details of the geology. The Pliocene volcanics

Panamint Butte as seen looking east from Father Crowley Vista Point. Precambrian–Permian sedimentary rocks deformed by Jurassic–Cretaceous thrusting and intrusion are overprinted by Miocene–Recent extension and covered by a blanket of basalt.

lie over brightly colored folded and altered Paleozoic sedimentary rocks, including marble and hornfels. There is also a band of light-colored Mesozoic granite that stretches across the canyon. The magma that formed the granite was the probable source of the heat that altered the Paleozoic rocks.

Rainbow Canyon had come to be called "Star Wars Canyon" by visitors who came to observe and photograph the US Air Force jets that trained by flying down the narrow canyon, just as Luke Skywalker and the Rebel pilots did in the Death Star trench in *Return of the Jedi*. Unfortunately, a jet crashed in the canyon in 2019, killing the pilot and injuring several visitors. That was the end of training flights through the canyon.

There are several long north–south valleys in Death Valley National Park. Looking east from Father Crowley Vista Point, across the Panamint Valley, I can see Panamint

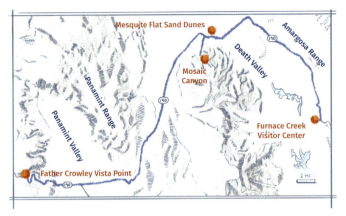

Map of the route from the Father Crowley Vista Point to Furnace Creek, Death Valley

Dunes, the Panamint Valley dry lake, and Panamint Butte on the horizon. Panamint Butte contains 15,000 feet of Paleozoic sediments, mostly dolomite but also including limestone, quartzite, and shale, that have been injected by 180-million-year-old Middle Jurassic granite. The sediments were folded and thrust-faulted during Late Jurassic and Cretaceous time, and normal-faulted over the past 15 million years; they are now covered by dark Pliocene basalt flows.

The road east from Father Crowley Vista Point snakes down the plateau to the valley below. Once on the valley floor, and continuing east on the now arrow-straight highway, I traverse the broad, flat, seemingly lifeless Panamint Valley. The only plants are widely scattered, hardy creosote brush. Helpful signs along the road warn me to turn off the air conditioning for the next ten miles to avoid overheating my engine. I'm glad I'm here during the cool season. As I enter the Panamint Range, tilted layers of coarse, boulder-filled gravel and alluvium alongside the highway are so fresh, they look for all the world as though they were eroded yesterday. I wend my way over the bleak and barren, east-tilted block of the Panamint Range, and enter Death Valley proper.

To get to Mosaic Canyon from the Father Crowley Vista Point, drive 40.5 miles (56 min) east on CA-190 to the unpaved Mosaic Canyon Road on the right; turn right (south) and drive to the Mosaic Canyon Trail parking area (36.572187, −117.144576). The turnoff to Mosaic Canyon is just west of Stovepipe Wells.

STOP 47: MOSAIC CANYON

Near the center of the national park, Mosaic Canyon on a late-winter day is a pleasant walk up a dry riverbed. I park and hike a quarter mile into these low hills, up this canyon to the narrows. Here I find water-polished walls consisting of angular rock fragments, a natural **terrazzo** chock-full of rounded and broken rock bits of all shapes and sizes. This is the Mosaic Breccia, *breccia* being the Italian word for "rubble." It is the lower portion of the Noonday Dolomite Formation. The fragments in the breccia were derived from broken parts of the lower Noonday Dolomite and from conglomerate eroded out of the underlying Kingston Peak Formation. The breccia has been interpreted as debris that accumulated at the base of a marine slope below limestone shoals that may have looked similar to the modern-day Florida Keys, except that this dolomite was created by ancient algae instead of coral. The Mosaic Breccia gives the canyon its name.

I continue another 1½ miles up-canyon until the way is blocked by a dry waterfall. Although I didn't see any, I'm told that there is scattered mining equipment in and around the canyon, left behind by prospectors during the late 1800s.

Small pools of standing water in rocky basins of the streambed remind me that Mosaic Canyon is a **slot canyon** carved by rare but powerful desert flash floods. It would be wise to check the weather before hiking into the canyon: flash floods are known to ream out canyons many miles downstream from where the rain actually falls.

In places this canyon is narrow enough that I can reach out and touch the canyon walls on both sides. The watercourse slices through Precambrian marble of the Noonday

Dolomite. The main body of the Noonday Dolomite is thought to have originated as a marine limestone that was altered to dolomite much later by the introduction of magnesium in the groundwater. It was later buried and altered to marble by the heat and pressure of deep burial.

There has certainly been enough time to effect such a metamorphosis. The dolomite is 640 million years old, give or take. This was a time before the seashells and corals that make up most limestones. How could a limestone or dolomite form that long ago?

As Noonday Dolomite contains abundant structures of algal origin, it was probably originally a shallow-water limestone. The evidence of algae is fossil slime—muddy mats of photosynthetic **cyanobacteria**, formerly called blue-green algae. These very thinly layered rocks are found today in the intertidal zone, areas of shallow water that are periodically exposed to the air. In marine environments, photosynthesis by cyanobacteria is the process that precipitates limestone. The limestone is formed when the bacteria inhales and removes CO_2 from the water during photosynthesis. In other words, the ancient limestone was formed by microbe breath.

This slot canyon reminds me that desert landforms, with very few exceptions, are the result of the presence and movement of water. As counterintuitive as that may sound in the middle of the driest place in North America, the fact is that once every hundred years, or maybe once in a thousand, there is a cosmic convergence of all the right factors, and the desert experiences a "perfect storm." When that happens, and several inches of water hit the parched landscape in just a few hours, the ground is simply unable to absorb that amount of water. What isn't immediately soaked up runs off across the flats or into canyons. A downpour can occur beneath a thunderhead, and the runoff will rush for miles as a surging torrent through all the low places.

When that happens, you do not want to be in a low place. The first thing you'll notice is a vibration in the ground, followed quickly by an audible rumble. The forerunner of the flood will be a slug of brown water, more mud than water, maybe only a few inches high, running and snaking its way down the canyon floor or a dry channel. If the downpour was far enough away, this may be your only warning of what is to come. This muddy slick will quickly be followed by a foaming flood coursing down the canyon carrying rocks and boulders the size of houses pried loose from canyon walls, a muddy slurry the consistency of thick hot chocolate that carries away everything in its path. The dense mixture is a **debris flow**, and you cannot outrun it. With the first warning, your best and only bet is to climb the canyon walls as fast as you can, because the canyon is about to be scoured out. If you ever wondered how a barn-sized rock got to the middle of desert flats, it was by a flash flood. These floods built up the alluvial fans, those piles of rock and sand that spread fan-like from the mouths of all desert canyons. It is these flash floods that drop boulders the size of a Greyhound bus at the mouth of every tributary canyon. Every other type of erosion pales in comparison to the overwhelming power of a flash flood.

Mosaic Canyon is a spectacular and colorful dry slot canyon.

When the violence is done, and the water relaxes and spreads across the center of a basin, new life is formed from the formerly churning chaos. Frogs, hiding beneath the surface in a kind of suspended animation, burrow their way upward and mate. Flower seeds, long dormant and waiting for this moment, burst into bloom. Desert pupfish in their isolated pools get a chance to spread out and find new holes to survive in. Bighorn sheep come down to drink.

Ancient humans were here too. Petroglyphs, Indigenous art made by scratching animal and geometric shapes into the rock surface, can be seen on the higher walls of Mosaic Canyon. If you encounter any of these artworks, or artifacts of any kind, please treat them with respect.

The other agent that affects desert landforms is wind. A short drive east takes me through Stovepipe Wells to the scenic dune field at Mesquite Flats.

To get from Mosaic Canyon to the Mesquite Flats Sand Dunes, return north to CA-190 and turn right (east); drive 4.5 mi (15 min) east to Mesquite Flats Sand Dunes parking area on the left (36.606174, −117.115049).

STOP 48: MESQUITE FLATS SAND DUNES

Mesquite Flats Sand Dunes is one of the few areas in North America with large, well-developed sand dunes. These active dunes are made up of quartz and feldspar grains blown in from the Cottonwood Mountains to the north and northwest. The grains accumulated in the valley bottom, having been carried by strong local winds. The tallest dunes are about 100 feet high, although both the height and location of the dunes shift daily with the winds.

Climbing a dune, even a modest one, is more work than you'd think when you look at it. The cliché "Two steps forward, one step back" actually applies in the loose sand. On the other hand, sliding down the face of a dune is a heck of a

Basin-and-Range 231

Mesquite Flats Sand Dunes.
The Funeral Mountains in the background consist
of Cambrian–Devonian sedimentary rocks and
Miocene–Recent sedimentary rocks and volcanics.

lot of fun. Just expect to get sand in your shoes, and probably everywhere else.

To get from the Mesquite Flats Sand Dunes to Furnace Creek, drive 22.8 miles (25 min) southeast on CA-190 to the Furnace Creek Visitor Center on the right (36.461071, −116.866270).

STOP 49: FURNACE CREEK VISITOR CENTER

Out of the frying pan, into the fire. As I mentioned earlier, Furnace Creek holds the record for the highest temperature ever recorded. I cannot think of a less hospitable place to live. And yet 136 people called Furnace Creek home in 2020. Located 190 feet *below* sea level, it is the headquarters for the national park, and since 1936 has been the Death Valley Indian Community reservation for the Timbisha Shoshone band, a group whose ancestral home included both Death Valley and the Owens Valley. Natural springs created an oasis

Furnace Creek, 1871 (Photo by Timothy H. O'Sullivan [1840–1882], Smithsonian Institution, Collection of the National Archives at College Park)

at Furnace Creek, but the springs have gradually faded away due to drawdown of the water table to support the village.

Originally called Greenland Ranch, after the fields of alfalfa planted here, the community was established in 1883 by the William Tell Coleman Borax Company. Greenland Ranch was renamed Furnace Creek Ranch in 1933. Furnace Creek was formerly the center of mining operations for the Pacific Coast Borax Company and its historic twenty-mule teams that hauled wagons of borax across the Mojave Desert.

The national park visitor center at Furnace Creek has a useful and informative raised-relief map of Death Valley as well as displays describing the geology, climate, and mining history of the area. Zabriskie Point lies a handful of miles southeast of Furnace Creek. It is known for its colorful badlands.

To get from Furnace Creek to Zabriskie Point, drive 5 miles (9 min) southeast on CA-190 to the Zabriskie Point parking area on the right (36.420967, −116.809300).

Basin-and-Range 233

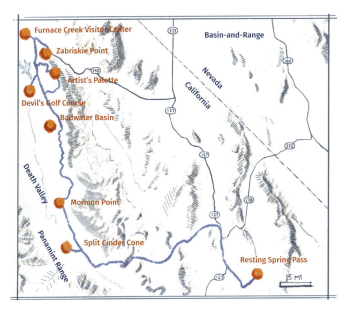

Map of the route from Furnace Creek to Resting Spring Pass

STOP 50: ZABRISKIE POINT

I time my arrival at Zabriskie Point just as the sun is going down. The lengthening rays of sunlight turn golden, and the yellow light seems to saturate the rocks with an intense glow. As the sun sets, the eroded, tortured, intensely gullied and barren landscape turns from shades of blond to pink to a dark crimson. This alone is worth the visit.

The multihued badlands at Zabriskie Point are striking even in full daylight. They consist of easily eroded shale, gravel, and volcanic ash, the remains of old lake deposits. This visual cacophony of stone belongs to the Miocene-age Artist Drive Formation, approximately 5–25 million years old. The prominent point to the west is known as Manley Beacon; the Black Mountains to the southwest are topped by dark basalt, from which they derive their name. The drab-brown layers above the Artist Drive Formation are full of cobbles and pebbles; they are part of a conglomeratic sandstone of the Pliocene-age Furnace Creek Formation, about 2.6–5 million

Manley Beacon and Artist Drive Formation at Zabriskie Point. The colors are especially stunning at sunset.

years old. These are the prominent coarse gravels I drove through in Furnace Creek Wash on the way here. The gravels were eroded from the nearby mountains. To the east are Precambrian and Paleozoic rocks of the Funeral Mountains. They have been raised up along the Furnace Creek Fault zone.

The following day, I encounter the same formations on the west side of the Black Mountains, also making a colorful splash, at Artist's Palette.

To get to Artist's Palette from Zabriskie Point, backtrack 3.5 miles north on CA-190 to Badwater Road; turn left (south) on Badwater Road and drive 13.3 miles (21 min) to Artist's Drive on the left; turn left and drive to Artist's Palette (36.363485, −116.802958).

STOP 51: ARTIST'S PALETTE

Artist's Drive is a one-way road that climbs up an alluvial fan, the rocky debris washed out of the mountains by flash floods. I'm driving east to the mountain front, where a sudden steepening of the slope indicates the scarp of the

Death Valley Fault. This is a near-vertical normal fault where the east side has moved up and tilted the rocks to the east, while the west side has dropped down into the valley. Like the San Andreas Fault system, the west side of this fault is moving slightly northwest, absorbing some of the movement of the Pacific Plate relative to the North American Plate.

From the high point on this road, I have a panoramic view of Death Valley laid out before me to the west. As well, there is the colorful Artist Drive Formation to the east, almost 5,000 feet thick. This is the same formation I saw at Zabriskie Point, although some of the colors are different. The rusty-red, pink, and yellow colors at Artist's Palette are a result of oxidation, or rusting, of iron minerals in the rock. The white and green colors are altered volcanic ash and ashy lake-bed sediments. Each of these colors has darker and lighter patches, as if a rare cloud were hovering overhead. The volcanic ash and lake-bed sediments are soft, hardly worthy of the term rock, and are readily eroded. The result is this intensely dissected, densely gullied hillside. The gaudy colors look as though they have been splashed seemingly at random across a giant Jackson Pollock canvas. I want to spend more time exploring here, picking up and touching the colors. But there is more to see.

Reluctantly I leave behind the rugged painted foothills and head out into the flats in the center of the valley.

To get from Artist's Palette to Devil's Golf Course, return to Badwater Road; turn left (south) and drive to West Side Road; turn right (west) and drive to Devil's Golf Course (36.339121, −116.867698), a total of 8.4 miles (20 min).

STOP 52: DEVIL'S GOLF COURSE

Full disclosure: I am *not* a golfer. I tried it a couple of times, and the word *mediocre* would be a compliment. But I know what a golf course looks like, and this ain't what comes to

*Broken salt crust at Devil's Golf Course.
View west to the Panamint Range.*

mind. The flat valley floor is a broken rubble of salt crust that is difficult to walk across without twisting an ankle. You would never find your ball if you tried to golf here. There is not a hint of green: everything is brown sand and white salt. On the other hand, the rugged and barren mountains that frame the valley are quite enchanting, especially in winter with a dusting of snow on the peaks.

The salt flats that make up Devil's Golf Course formed when Lake Manly, the ice age lake that filled this valley, dried up. Minerals dissolved in the lake water precipitated a salt crust that can be up to 5 feet thick. Salt-rich groundwater rises from the underlying mud, drawn upward by **capillary action**, an upward suction caused by evaporation at the surface. As the brine dries, it creates salty pinnacles that rise out of the desert floor. The pinnacles grow slowly, perhaps half an inch in 15 years. The dried and cracked salt slabs, like thick **mudcracks**, jostle with each other as the soft mud below shifts ever so slightly.

A short drive south along the valley floor brings us to the lowest point in the valley, and the lowest elevation in North America.

To get from Devil's Golf Course to Badwater Basin, return to Badwater Road and turn right (south); drive 10.6 miles (16 min) south on Badwater Road to the Badwater Basin turnout on the right (36.229793, −116.767285).

STOP 53: BADWATER BASIN

I am now 282 feet below sea level. What is truly strange is that just 85 miles west of me is the highest point in the conterminous United States, Mount Whitney. The questions that pop into my mind are "How did it get so low?" and "Why isn't the ocean filling it?"

As mentioned earlier, the land began to split apart here when the San Andreas Fault system extended into this area around 15 million years ago. This led to the development of sideways-moving fractures in Death Valley. These ruptures include the Furnace Creek and Sheephead Faults that bound the valley here. As they moved, they caused the area between them to drop. This particular type of basin has been called a **rhombochasm** due to the squeezed rectangle shape of the basin. In other words, as the Earth's crust pulled apart along the bounding faults, the bottom simply dropped out of the valley.

The valley is roughly 12 miles wide by 80 miles long, but it is bounded by faults and mountains on all sides, so there is no connection to the Gulf of California almost 300 miles to the south. Like the Salton Trough further south along the San Andreas Fault (226 feet below sea level), this is a valley with a floor below sea level that is cut off from the sea by a dam of uplifted earth and rock.

The valley was full of water, though, during the previous ice age. The valley floor is covered by a salty layer 3–6 feet thick, the remains of the same Lake Manly that evaporated at Devil's

George Grant, the first photographer for the National Park Service, took this photograph in Death Valley at Badwater in 1935. The elevation has since been resurveyed and found to be 282 feet below sea level. (National Park Service Photo/George A. Grant)

Golf Course. Unlike Devil's Golf Course, Badwater Basin does not have salt pinnacles and crust. Because it is slightly lower and floods during rare rain events, any pinnacles that might have formed are dissolved in the stagnant brine. There is shallow water in winter and amazing giant polygonal mudcracks in the drying salt flats the rest of the year.

And the water is bad. This briny soup is full of concentrated minerals and salt. Do not attempt to drink it.

Separating Badwater Basin and the Black Mountains to the east is the steeply inclined Death Valley Fault. The Black Mountains are made up of a 1.7-billion-year-old gneiss; the mountains were intruded fairly recently by gabbro, a dark magma with the same composition as basalt. The gabbro belongs to the Willow Spring pluton, around 11 million years old. It erupted to form the black basalt that flowed over the top of the range. Since that time, the mountains were uplifted as the valley was dropped down. As the range rose, the overlying sedimentary rocks slid off on low-angle normal faults. This left the dome-shaped core of the mountains exposed, the Badwater Turtleback. This is the northernmost of three such features exposed in the Black Mountains. At my next stop, I'll be looking at the Copper Canyon Turtleback just north of Mormon Point.

To get from Badwater Basin to Mormon Point, drive 16 miles (20 min) south on Badwater Road to Mormon Point on the right shoulder (36.058473, −116.764617). This stop is actually a pullout half a mile north of the Mormon Point marker.

STOP 54: MORMON POINT AND THE COPPER CANYON TURTLEBACK

Looking north from Mormon Point along the front of the Black Mountains, I can easily see a large hump-shaped mountain, the Copper Canyon Turtleback. The undulating surface of the mountain is a shallow fault that may have been caused by gravity-sliding of the overlying rock off the top of the mountain. It is thought that the Black Mountain Fault marks the zone along which the top of the mountain, made of unmetamorphosed, brittle sedimentary rock, slid west into Death Valley as the mountains were uplifted. The surface of the turtleback is a thin veneer of highly sheared carbonate rock. Geologists who study rock mechanics know

Looking northeast to the Copper Canyon Turtleback structure from just north of Mormon Point

that carbonates such as limestones can flow under pressure. It is as if these rocks were the grease that let the upper part of the mountain slide down the slope into the deepening valley.

So far, I've seen mostly normal faults, where one side goes up and the other goes down. Some of these, like the Death Valley Fault, had a component of sideways slip, but it isn't obvious. At the next stop, sideways motion is dominant—because it literally split a mountain.

To get from Mormon Point to Split Cinder Cone, continue driving south on Badwater Road to West Side Road; turn right (west) on West Side Road (graded, but unpaved) and drive to Split Cinder Cone (35.943471, −116.734372), for a total of 11.1 miles (21 min).

STOP 55: SPLIT CINDER CONE

The Death Valley Fault has been active for the past 3 million years or so. Around 300,000 years ago, basaltic magma rose to the surface along the fault zone. When it reached the surface,

Basin-and-Range 241

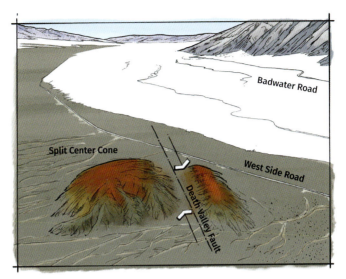

Split Cinder Cone is severed by the Death Valley Fault. This is an oblique view north. West Side Road passes by the mountain.

gas entrained in the lava expanded and blasted the molten rock into the air. It fell as cinders, forming a cone-shaped volcanic mound, a cinder cone. After the eruption ended, the fault continued to move. The opposite side of the fault moved right, making this a **right-lateral slip fault**.

If you could look down on the cinder cone, you would see that the fault moved one side of the small volcano to the southeast and the other side to the northwest. (This is the same type of offset seen on all strands of the San Andreas Fault system.) Since the eruption, the two sides have been separated by 300 feet. That works out to an average movement of a foot every thousand years.

I leave Death Valley now, on my way to Las Vegas. About 15 miles before the Nevada border there is an amazing roadcut that summarizes the Basin-and-Range in miniature.

To get from Split Cinder Cone to Resting Spring Pass, return to Badwater Road and turn right (south); drive southeast on Badwater Road to CA-178/Jubilee Pass Road; turn left (east) on

CA-178 and drive to CA-127; turn right (south) on CA-127 and drive to CA-178; turn left (east) onto CA-178 and drive to Resting Spring Pass pullout on the right (35.997167, −116.219723), for a total of 36.2 miles (35 min).

STOP 56: RESTING SPRING PASS

This dramatic outcrop might be considered a small-scale version of the Basin-and-Range province, in that there are multiple normal faults that bound tilted mountain blocks. You can clearly see the same layers offset on either side of the faults, like a stack of tilted dominoes. The outcrop is mainly pink-and-white Miocene-age Resting Spring Pass tuff. This welded volcanic ash is 200–800 feet thick in the Resting Spring Range. Despite the widespread distribution of this tuff, the source is unknown and probably buried beneath a nearby valley. That the 12-million-year-old ash is tilted indicates clearly that it is older than the faulting that tilted it, and probably older than the faulting that created the range itself.

The Resting Spring Range is an east-tilted, faulted mountain range composed mainly of very old, Cambrian-age strata. The layers of all succeeding ages were deposited, then eroded, and finally the Cambrian rocks were covered by much younger volcanics on the gentle eastern slope of the range.

The roadcut has a vivid black slash that cuts diagonally across the outcrop. This is a 9.5-million-year-old obsidian layer within the welded tuff. The presence of this volcanic glass indicates rapid quenching of a silica-rich magma.

Normal faults are well exposed in the roadcut. The faults offset both the tuff and layers of the Tertiary-age Chicago Valley Formation that lies below the tuff. From bottom to top, the Chicago Valley Formation consists of conglomerate, tuff, and limestone. This sequence indicates an upward progression from alluvial valley fill deposits to volcanic activity, and finally to a clearwater lake. The normal faults are inclined both to the northwest and to the southeast. The amount of offset along the faults ranges from a few inches to a few feet.

Basin-and-Range   243

Resting Spring Pass roadcut through faulted Resting Spring Pass tuff

Whereas overall Basin-and-Range extension is roughly east–west, the direction of extension in the Resting Spring Pass area, based on fault orientations, is northwest–southeast. The direction that an area is extending is almost always perpendicular to the fault direction: think of pulling apart two rectangular block magnets. Which just illustrates that you can have local variations within the overall east–west-extending Basin-and-Range.

I'm getting out of my vehicle to have a closer look at this classic geological roadcut when a car, a rare sight on this remote stretch of highway, slows down to see if anything is wrong. "Need a hand fixing a flat?" the man asks. "Nah, I'm just looking at rocks." He looks at me like I'm nuts and, as the car speeds away, yells "Good luck to ya." And it hits me: this is what I love about California. It doesn't matter how crazy you might appear; someone will give you the space to do your thing. Mark Twain is claimed to have said that "America is built on a tilt, and everything loose slides to California."[13] I feel like I'm finally home.

Epilogue

What began as a land of abundance, with diverse peoples and landscapes teeming with game and food plants, fragmented by geology, shifted rather suddenly after the appearance of Europeans. The discovery of gold at Sutter's Mill in 1848 transformed California from a sleepy Spanish/Mexican cattle-raising backwater to eventually become the fifth-largest economy in the world. News of the Gold Rush drew thousands of fortune seekers from around the world. The influx irrevocably altered California's economy and population. Soon there were bustling mines and farms and railroads everywhere. The state sprouted oil fields to the extent that it became a significant contributor to the global hydrocarbon industry. The black gold bubbling beneath the surface played a pivotal role in California's modern economic landscape. The collision of tectonic plates led to the formation of other valuable deposits: free steam to turn electrical turbines, and borax to clean your laundry. The state's world-class wine region owes its success to a combination of soil types, climate, and geological features. The Napa Valley thrives on the mixed remnants of ancient seafloor and volcanic activity, creating an ideal terroir for wine. The soil of the Central Valley, shed from the rising Sierra Nevada into an ancient seaway, is among the most productive in the world, hitting well above its weight. The Central Valley alone produces 8 percent of America's food supply from just 1 percent of the country's farmland. Half of all the fruits and vegetables grown in the US come from California. The geological wealth of California lies in its rocks and soils, in its precious metals and the energy that powers our world. The story of how the Earth's riches have shaped the state we know today is captivating.

Geology will also determine the state's future. To quote a saying that's sometimes attributed to Ibn Khaldun, and later to Napoleon Bonaparte, "Geography is destiny." Insofar as geology has determined the state's geography, together they will shape the state's destiny. From solar farms in the desert to lithium extraction from the Salton Sea and rare earth mining at Mountain Pass, California has the minerals and technology and innovative spirit to harness the future.

I set out to get reacquainted with the state I was born and raised in. This journey has taken me across California from northwest to southeast, from foggy, forest-shrouded coast to barren, dry desert. I've seen the San Andreas Fault at the edge of the North American tectonic plate, subduction zones, geyser fields, petrified forests, world-class wine country, oil fields and gold fields, glacially carved mountains and valleys, the remnants of glacial lakes, volcanic flows and columns, hot springs, narrow defiles cut by flash floods, badlands, sand dunes, and the hottest, driest, and lowest point in North America. There are not many places on Earth where you can see that much geology in such a compact area.

My next stop is Las Vegas, about 75 miles (1 hr 15 min) to the east. Not to gamble or see a show, mind you. Rather, to see a classic example of a thrust fault, the Keystone Thrust, in Red Rock Canyon. And to get a first glimpse of the Great Unconformity (a gap in the rock record that spans 1.2 billion years) and the rocks that make up the Grand Canyon: these are exposed at Frenchman Mountain on the east side of town. And to see the flaming, swirling sandstone at Valley of Fire State Park. Las Vegas is the gateway to a whole different geologic province, the Colorado Plateau. The Colorado Plateau province extends roughly from St. George, Utah, to Grand Junction, Colorado, and includes such cathedrals of nature as the Grand Canyon and Monument Valley, Zion Canyon and Arches National Parks, the Grand Staircase, and Canyonlands. It is a land of extraordinary beauty carved by snowmelt streams into colorful stone, of crimson cliffs and bountiful buttes, of slot canyons and steep gorges. A land brimming with national parks and monuments, arguably one of the most scenic corners of the globe. But all that is for another time.

Acknowledgments

I thank my son Adam for preparing the line drawings that appear throughout this book, and my editors at Heyday, Emmerich Anklam, Marthine Satris, and Michele Jones, for helpful advice and suggestions that made this a much better, much easier book to read. Thanks to Ron Kaufman for setting me on this path, and to Tom Killion for allowing me to use his amazing prints to help illustrate the text. Thanks to my son Ben for helpful suggestions on how to write a narrative, and to Ben and Melissa for accompanying me on parts of this journey. I especially wish to thank Steve Wasserman, the publisher at Heyday, for bringing this project to my attention and encouraging me to continue despite several false starts. Finally, I thank my wife, Nancy, for taking the wheel (both literally and figuratively) whenever I got carried away by the rocks.

Glossary

(because not everyone is a geologist . . .)

actinolite: a green mineral that is a calcium-magnesium-iron silicate.

active margin: a tectonically active continental margin, such as the west coasts of North and South America. This transition between ocean and continental crust is characterized by active faulting, volcanism, and/or mountain-building.

alluvial fan: triangle-shaped piles of gravel and sand that spread fan-like from the mouths of canyons.

alluvium: clay, silt, sand, and gravel deposited by flowing streams in a river valley or delta. Alluvial valley fill is the material filling a valley above the bedrock, or basement rock.

altered rock: a rock that has undergone chemical or mineral changes after it was deposited or emplaced.

ammonites: extinct spiral-shelled marine cephalopods. They are related to a living group that includes octopuses, cuttlefish, and squid. The earliest ammonoids appeared during the Devonian, with the last vanishing at the end of the Cretaceous.

amphibole: a family of dark silicate minerals that occurs in igneous or metamorphic rocks, notably hornblende, tremolite, and actinolite.

andesite: a medium-dark eruptive igneous rock, intermediate in composition between a rhyolite and basalt. It is typical of the Andes, for which it was named.

anticline: an arch-shaped fold in layered rock.

Antler Orogeny: a mountain-building event, in Nevada and adjacent areas, that began in the Late Devonian and continued into Early Pennsylvanian time.

arête: a sharp, narrow ridge of rock that separates two valleys, often caused by glacial erosion.

argillite: a hard, slaty, fine-grained sedimentary rock composed mainly of clay particles.

auriferous gravel: gold-bearing gravel.

badlands: large tracts of heavily eroded, barren land.

barrel of oil: a unit of volume used for measuring oil, defined as exactly 42 US gallons.

basalt: a black, fine-grained eruptive igneous rock rich in iron and magnesium. It flows readily, and forms much of the ocean floor.

basement rock: igneous or metamorphic rocks lying beneath all other layered rocks and sediments.

basement-cored vertical uplift: basement rock and overlying sedimentary rocks that have been uplifted by vertical or near-vertical faults.

Basin-and-Range: a geological province extending from eastern California to central Utah and from Oregon south into northern Mexico. It is characterized by abrupt changes in elevation, alternating between narrow faulted mountain ranges and long flat-bottomed valleys or basins. The province is a result of tectonic extension that began around 17 million years ago in the Miocene.

batholith: a large mass of intrusive igneous rock, larger than 40 square miles in area, that forms from cooled magma deep in Earth's crust. Batholiths are almost always made of granite or granodiorite.

biotite: a black silicate mineral in the mica group.

blueschist: a bluish rock formed by metamorphism of basalt under conditions of high pressure and relatively low temperature, found at depths of 9–18 miles.

borax: a powdery white substance, also known as sodium borate, composed of boron, sodium, and oxygen. It's widely used as a household cleaner and laundry detergent booster. It is generally found in dry lakebeds.

brachiopods: a phylum of marine animals that have two shells; it includes, for example, the pecten, which most people recognize as the symbol for Shell Oil Company.

breccia: a rock consisting of angular fragments cemented together.

bull quartz: milky-white quartz occurring in veins and sometimes containing gold.

caldera: the crater left behind after a large volcanic eruption, where so much lava and ash are ejected that the empty magma chamber collapses in on itself.

capillary action: the process of a liquid flowing upward in a narrow space due to the forces of adhesion, cohesion, and surface tension, and in opposition to the downward pull of gravity.

Cascade volcanic arc (the Cascade Range): a major range of volcanoes in southwestern British Columbia, Washington, Oregon, and Northern California that is a result of subduction and melting of the Juan de Fuca Plate beneath the North American Plate.

Central Valley: see Great Valley.

chalcopyrite: a brassy golden-yellow mineral that is a copper-iron sulfide, an ore of copper and, to a lesser extent, iron, both of which are bound up with sulfur.

chert: a hard, opaque rock composed of silica with microscopic quartz crystals. It occurs as nodules (flint) or as layers.

chlorite: a green clay mineral that is a hydrated aluminum silicate common in low-grade metamorphic rocks and altered igneous rocks.

cinder cone: a steep conical hill of loose volcanic fragments, such as ash or pumice, blown out of a vent.

cirque: an amphitheater-like bowl formed at the head of a valley by glacial erosion.

col: the lowest point on a ridge or saddle between two peaks, typically a pass.

columnar basalt: columns of black basalt lava formed as the lava cooled, contracted, and cracked.

columnar jointing: cracks, caused by cooling and hardening of lava, that form hexagonal or pentagonal columns of rock.

conglomerate: sedimentary rock consisting of rounded cobbles and pebbles.

continental crust: mainly granitic rocks that form the Earth's continents.

continental margin: the edge of the continent, the area just before it transitions from continent to seafloor.

continental shelf: that part of the continent that extends under the sea to about 200 meters (650 ft) in depth.

continental slope: that part of the continent that extends from the outer edge of the continental shelf to the deep ocean floor.

convection current: a heat-driven currents of molten rock in the Earth's mantle, caused by a difference in temperature between the Earth's interior and the crust. The current transfers heat from the interior to the surface of the Earth at oceanic spreading centers, then cycles cool rock back down into the mantle at subduction zones.

country rock: the native rock in an area.

crust: the outer 12–50 miles of the Earth, thicker on the continents and thinner in the oceans.

crystal system: categories of crystals classified according to the geometry of the crystal axes.

cyanobacteria: also called blue-green algae, these

microscopic organisms are found in all types of water and use chlorophyll, in a process called photosynthesis, to convert sunlight into metabolic energy.

dacite: a medium-gray volcanic rock that is high in silica and is intermediate in composition between andesite and rhyolite. It is composed predominantly of plagioclase feldspar and quartz.

debris flow: a water-laden mass of soil and broken rock that flows down mountainsides, funnels into streams, and forms thick, muddy deposits on valley floors.

dike: a vertical or near-vertical sheet-like magma intrusion.

diorite: an igneous rock that cools in the subsurface and is composed mainly of the silicate minerals plagioclase feldspar, biotite, and hornblende. The chemical composition of diorite is intermediate between that of basalt and granite.

dolomite: both a mineral composed of calcium-magnesium carbonate, and a rock type similar to limestone that consists mainly of the mineral dolomite.

drainage divide: a ridgeline or elevated terrain that separates neighboring watersheds.

East Bay Hills: also known as the Berkeley Hills or Oakland Hills, this is a range in the Pacific Coast Ranges on the east side of San Francisco Bay.

East Pacific Rise: a mid-ocean spreading center that is also a tectonic plate boundary. It is located along the floor of the Pacific Ocean more or less parallel to the west coasts of South and Central America. It's northern termination, in the Gulf of California, transitions into the San Andreas Fault.

electrum: a naturally occurring alloy of gold and silver, with trace amounts of copper and other metals.

escarpment: a long, steep slope, especially along a fault, that separates land at different heights.

exfoliation: a form of weathering in which curved plates of rock are stripped from the rock underneath due to pressure release. This results in exfoliation domes and rounded boulders. Exfoliation domes occur because of joints that are curved essentially parallel to the ground surface.

Farallon Plate: an ancient oceanic plate east of the Pacific spreading center. It began subducting under the North American Plate during the Early Jurassic. Over time, the central part of the Farallon Plate was completely subducted beneath the North American Plate. The remains of the Farallon Plate are the Explorer, Gorda, and Juan de Fuca Plates subducting under North America; the Cocos Plate subducting under Central America; and the Nazca Plate subducting under South America.

fault: a more or less planar rock fracture in which the two sides move in opposite directions.

feeder dike: a near-vertical sheetlike igneous body that feeds magma to volcanic vents at the surface.

feldspar: an alumino-silicate mineral containing varying amounts of potassium, sodium, calcium, and/or lithium. The cream to orange-pink feldspar group is the most common group of minerals in the Earth's crust, making up about 50 percent of all rocks.

fissure: a long, narrow opening made by cracking or splitting of rocks.

flat-slab subduction: subduction that is characterized by a low subduction angle (less than 30 degrees from horizontal) for the descending plate as it moves beneath the continent.

fold: a wave-like structure that forms when rock layers bend instead of breaking during deformation.

formation: a group of related rock layers that is distinct enough to be mapped as a unit—for example, layers of sandstone sandwiched between layers of limestone. A formation is usually named after the location where it is best exposed or where it was first described.

Franciscan Terrane (also Franciscan Complex, Assemblage, or Mélange): a Jurassic through Cretaceous group of rocks found in the California Coast Ranges. The Franciscan is dominated by graywacke sandstones, shales, ribbon cherts, pillow basalt, and serpentinite. Other rocks include conglomerates, limestones, flow basalts, and basalt dikes. All of these rocks have experienced metamorphism, some of it gentle, some intense.

Franciscan trilogy: the sequence, from bottom to top, of pillow basalts, ribbon cherts, and turbidites/graywacke sandstones.

fumarole: a vent at the surface where volcanic gases and vapors are emitted. Fumaroles are an important sign that a volcano is active.

galena: a lead sulfide mineral that often contains a significant amount of silver.

gastropods: a class of mollusca that includes snails and slugs, among others.

geothermal energy: electricity produced by utilizing heat from within the Earth to generate steam and turn turbines. Geothermal wells are drilled to capture the Earth's heat.

geyser: a hot spring that is under pressure and erupts hot water and steam.

glacial polish: the mirror-like finish on rocks that have been polished to a sheen by glaciers moving over their surface.

glacier: an accumulation of ice and rock that moves downslope under the influence of gravity.

glauconite: a green mineral in the mica group.

glaucophane: a blue-gray mineral characteristic of high pressure–low temperature metamorphism.

gneiss: a coarsely crystalline banded metamorphic

rock indicative of high temperature and pressure.

granite: a light-pink, coarsely crystalline igneous rock that cooled underground and is rich in quartz and feldspar.

granodiorite: a black-and-white coarsely crystalline igneous rock similar to granite.

graywacke: a dark, dirty sandstone.

Great Basin: an internally drained area in western North America including parts of Utah, Nevada, eastern California, southeast Oregon, and southern Idaho.

Great Valley: California's Central Valley, a broad, flat valley between the Sierra Nevada and Coast Ranges.

Great Valley Sequence: a 40,000-foot-thick group of geologic formations, of Jurassic and Cretaceous age, deposited in an ancient seaway that corresponds roughly to the current Central Valley of California.

hanging valley: a glacial valley that is a tributary to a larger glacial valley and ends in a drop to the main valley, usually with a waterfall.

headframe: a structure built above an underground mine shaft for hoisting and lowering machines, people, and materials into and out of the mine.

horn: a pointed peak that was carved on at least three sides by glaciers. The classic example is the Matterhorn.

hornblende: a black mineral common in igneous and metamorphic rocks.

hornfels: a finely crystalline metamorphic rock, usually dark and without a platy texture.

hydraulic mining: a form of mining that uses high-pressure jets of water to dislodge rock material or move sediment downslope to where it can be sluiced for gold nuggets.

hydrothermal fluids: hot groundwater, often acidic, that circulates in the Earth's crust and carries dissolved minerals and metals that precipitate

Glossary

with a drop in temperature or pressure. These hot waters can chemically alter surface rocks and minerals, and often emerge at hot springs.

ice age: a long period of cooler-than-normal Earth temperatures that results in continental ice sheets and glaciers.

ice sheet: a stationary mass of ice that covers a large area.

ignimbrite: a volcanic rock, dominated by hot gas, ash, and pumice fragments, from an explosive eruption that flows at high temperature over the countryside.

intrusive rock: igneous rock that cools far beneath the surface.

kettle: a water-filled depression formed by blocks of melting glacial ice left behind by a retreating glacier.

lakeshore terrace: the eroded former shoreline of an evaporating lake.

Laramide Orogeny: a period of mountain-building in western North America that began in the Late Cretaceous (70–80 million years ago) and ended in the Eocene (about 35–55 million years ago). The mountains contain granite cores and are bounded by steep faults. The compression that caused this event is thought to be related to shallow subduction of the Farallon Plate beneath the North American Plate.

lateral moraine: piles of rock and debris carried by and dropped along the margins of glaciers.

lens: a body of rock that is thick in the middle and thin on the ends.

lime: calcium carbonate, the main component of seashells.

limestone: a rock consisting mostly of seashells or calcite (calcium carbonate).

lode: a vein of metal ore.

magma chamber: the location below a volcano where magma—molten rock—is stored before an eruption.

magmatic arc: also called a volcanic arc, this is a belt of volcanoes formed above a subducting oceanic tectonic plate, with the volcanoes arranged in an arc shape.

magnitude (of an earthquake): a measure of the size, or amplitude, of the seismic waves generated by an earthquake.

mantle: the mostly solid bulk of Earth's interior. The mantle lies between Earth's core and its thin outer crust. The mantle is about 1,800 miles thick, and makes up 84 percent of Earth's total volume.

marble: a metamorphic rock consisting of carbonate minerals that have recrystallized under the influence of heat and pressure. The original rock is usually limestone.

marine terrace: a relatively flat surface of marine origin, usually eroded by waves, and bounded by steeper slopes.

mélange: rock characterized by a lack of layering and the inclusion of fragments of all sizes and rock types, contained in a fine-grained deformed matrix.

metamorphic core complex: exposures of deep crust brought to the surface by extension. During this process, metamorphic rocks that formed under extreme heat and pressure are uplifted and exposed beneath low-angle faults.

metamorphic rock: originally igneous, sedimentary, or even previously metamorphosed rock that is changed when subjected to heat, pressure, hot mineral-rich fluids, or some combination of these. The changes can be chemical or textural or both.

metamorphism: a change in the composition or fabric of a rock by heat, pressure, or hot fluids.

metasediment: metamorphic rock that was originally a sedimentary rock such as sandstone or mudstone.

metavolcanic: metamorphic rock that was originally a volcanic rock such as rhyolite or basalt.

mica: a family of platy minerals, ranging from white to black, that is common in igneous rocks.

mid-ocean ridge: a long, seismically active submarine ridge system marking the site of the upwelling of magma associated with seafloor spreading. Examples include the East Pacific Rise and the Mid-Atlantic Ridge.

Moment magnitude scale: a scale developed to capture all the different seismic waves from an earthquake no matter how close or far from the epicenter. The scale measures the size of the earthquake and is more accurate than the Richter scale for very large earthquakes.

moraine: an accumulation of loose rock debris that occurs in glaciated regions and that has been carried along and deposited by a glacier.

Morrison Orogeny: also called the Sonoma Orogeny, this Permian–Triassic mountain-building event occurred about 250 million years ago and is thought to represent a subduction event and the associated accretion of a volcanic island arc onto the western margin of Pangea (what would later become North America).

mudcrack: a sedimentary feature formed when muddy sediment dries and contracts. The cracks often link up to form five- or six-sided polygons, with the muddy slab inside the polygon curling up like a bowl.

Nazca Plate: an oceanic tectonic plate in the eastern Pacific Ocean off the west coast of South America. It is a remnant of the breakup of the Farallon Plate about 23 million years ago.

Nevadan Orogeny: a mountain-building event that occurred along the western margin of North America during the Late Jurassic to Early Cretaceous approximately 155 to 145 million years ago. The rocks of the Sierra Nevada were formed by the Nevadan Orogeny.

normal fault: a fault in which the block above the fault has moved down relative to the block below.

North American Plate: a tectonic plate containing most of North America, which extends from the Mid-Atlantic Ridge to the San Andreas Fault and East Pacific Rise. The plate includes both continental and oceanic crust.

nuée ardente: a French term for "glowing cloud," these are turbulent, fast-moving clouds of hot gas and ash erupted from a volcano. They can flow downslope at speeds that exceed 50 mph and temperatures of 400 °F–1300 °F. These materials collectively are called pyroclastic flows.

obsidian: a volcanic glass, usually black but sometimes reddish, that forms when lava with granitic composition cools quickly.

obsidian dome: a rhyolite dome consisting of obsidian (a volcanic glass) and pumice (gas-filled volcanic froth).

oceanic trench: a long, narrow topographic depression of the ocean floor that marks the location where tectonic plates move toward each other and one slab descends beneath another slab. Trenches are generally parallel to and about 120 miles from a volcanic arc.

olivine: a green magnesium-iron silicate mineral common in the Earth's mantle and in igneous rocks.

ophiolite: oceanic crust and upper-mantle rocks that have been thrust or uplifted and exposed at the surface, and are often emplaced on top of continental crust.

outcrop: rock exposed at the surface.

Pacific Plate: an oceanic tectonic plate that lies beneath the Pacific Ocean. The plate forms at the East Pacific Rise spreading center and generally moves northwest.

paleontologist: a person who studies the history of life on Earth through the fossil record.

Pangea: the supercontinent that existed during the late Paleozoic and early Mesozoic eras. It assembled from the earlier continents of Gondwana,

Euramerica, and Siberia approximately 335 million years ago, and began to break apart about 200 million years ago, at the end of the Triassic and beginning of the Jurassic.

passive margin: the transition between oceanic and continental crust formed on the margins of an ancient rift. Eventually the continental rift forms a mid-ocean ridge, and the continental margin, now quiet, moves away from the oceanic spreading center. This is in contrast to an active margin, where mountain-building is occurring.

pelecypods: clam and clam-like sea creatures.

peridot: the rare transparent and yellow-green variety of the common mineral olivine. Peridot forms deep in the Earth's mantle and is brought to the surface, usually by volcanoes.

peridotite: a dense, coarse-grained, olivine-rich intrusive igneous rock. It is a common component of oceanic crust that is derived from the upper mantle.

permeability: the ability of a material to allow liquids or gases to pass through it. It is a measure of the connectedness of holes and voids in a rock.

petrified: referring to organic matter such as bone or wood that has been replaced by minerals and turned to stone.

phyllite: originally a mudstone, now a foliated, finely layered metamorphic rock rich in mica.

pillow lava: lava, commonly basalt, consisting of characteristic pillow-shaped blobs attributed to extrusion of the lava underwater. It forms part of normal oceanic crust.

placer: a type of mineral deposit in which grains of a valuable mineral such as gold are deposited and concentrated by river currents or wave winnowing along a beach. Also a mining method that uses water and gravity to separate gold from surrounding material.

plankton: a diverse collection of organisms that float

and drift in water. The organisms provide a crucial source of food to many small and large aquatic organisms. Marine plankton includes single-cell bacteria, algae, protozoa, microscopic fungi, and drifting or floating animals.

plate tectonics: the theory that explains how major landforms are created as a result of Earth's movements. The theory explains many phenomena, such as mountain-building, volcanoes, and earthquakes.

playa: a dry lakebed.

pluton: a typically large body of intrusive igneous rock.

porosity: a measure of the void space in rock.

proto–North America: the continental crust of the supercontinent Pangea and its predecessors, which would later become North America.

pumice: a very light glass that is formed by the rapid cooling of lava from volcanoes; it is full of small gas bubbles and holes.

pyrite: a shiny yellow or golden iron sulfide mineral, also known as "fool's gold."

pyroclastic flow: a fast-moving current of hot gas and volcanic matter that flows along the ground away from a volcano at average speeds of 60 mph and that can reach temperatures of about 1800°F. These flows are a result of explosive eruptions.

pyroxene: a common dark-green to black iron-magnesium silicate mineral group found in igneous and metamorphic rocks.

quartz: silica (silicon dioxide), the most common mineral; it is usually clear, but can be white or other colors.

quartz monzonite: an intrusive igneous rock related to granites.

quartzite: a hard, usually white metamorphic rock that was originally a pure quartz sandstone.

radioactive decay: the process by which an unstable atomic nucleus loses energy by radiation.

radioactive isotope: an atom that has excess numbers of either neutrons or protons, giving it excess nuclear energy and making it unstable. This excess energy can be emitted from the nucleus as gamma radiation, transferred to one of its electrons to release it, or used to create and emit a new particle from the nucleus. This process is called radioactive decay.

radiolarians: single-celled protozoa that produce mineral skeletons usually made of silica. They are a free-floating marine plankton.

rhombochasm: a basin formed between overlapping strike-slip faults, roughly rhombohedral in shape. Such a basin forms when overlapping strike-slip faults create an area of crustal extension.

rhyolite: the most silica-rich of volcanic rocks, it is generally a fine-grained, white to pink rock with the same mineral composition as granite.

rhyolite dome: a circular, dome-shaped hill resulting from the slow extrusion of viscous rhyolite lava.

ribbon chert: thin and hard silica-rich chert beds separated by thin beds of soft, clay-rich shale.

rift: an elongated, linear valley between highlands or mountain ranges produced by the process of rifting. Rifts are formed as a result of the pulling apart and faulting of the Earth's crust due to extension.

right-lateral slip fault: a fault in which the displacement of the opposite side is to the right (when you are looking across the fault).

rôche moutonnées: a rock formation created by the passing of a glacier. The passage of glacial ice over underlying bedrock often results in an erosional form that is gently sloping on the upstream side and steep on the downstream side.

roof pendant: a mass of country rock that projects

downward into or is entirely surrounded by an igneous intrusion such as a batholith or pluton.

scarp: a steep bank, slope, or escarpment caused by vertical movement along a fault.

schist: a metamorphic rock with well-developed thin layering or foliation ("schistosity") produced by alignment of mineral grains under high pressure.

sediment core: sediment sample collected during drilling operations.

seismic survey: the use of reflected sound waves to produce a "CAT scan" of the Earth's subsurface.

seismically active: referring to an area with lots of earthquakes.

seismologist: an earth scientist or geophysicist who studies the origin and propagation of seismic waves in the Earth.

serpentine: a dark-green mineral consisting of hydrated magnesium silicate. It originates in the Earth's mantle.

serpentinite: a dark-green metamorphic rock composed of serpentine-group minerals.

Sevier Orogeny: a mountain-building event in western North America extending from northern Canada to Mexico. This orogeny was caused by the subduction of the oceanic Farallon Plate beneath the continental North American Plate between around 160 and 50 million years ago. The event was characterized by thrust faulting.

shale: a relatively soft, fine-grained sedimentary rock that was originally mud.

shear zone: a thin zone in the Earth's crust that has been strongly deformed due to the rocks on either side of the zone slipping past each other. In the upper crust, where rock is brittle, the shear zone takes the form of a fault.

sheeted dikes: more or less parallel intrusions of igneous rock, forming vertical layers in oceanic crust. At mid-ocean ridges, dikes form when magma rises into and fills fractures in areas of diverging tectonic plates.

Sierra Crest (also Sierra Divide): the ridge at the highest point of the Sierra Nevada that separates rivers and streams flowing west from those flowing east.

Sierra escarpment: the steep eastern slope of the Sierra Nevada that coincides with the Sierra Nevada Fault.

silica: the mineral silicon dioxide, found naturally as quartz.

siliceous ooze: marine sediment, rich in silica, derived from the silica shells of certain plankton such as diatoms and radiolarians.

slab rollback: when a section of subducted oceanic plate shifts from being nearly horizontal below continental crust to diving steeply into the mantle.

slag: stony waste matter separated from metals during the smelting or refining of ore.

slate: fine-grained thinly layered metamorphic rock derived from shale.

slot canyon: a long, narrow, sometimes winding canyon with sheer rock walls that are typically eroded into sandstone; these canyons are prone to flash flooding.

sluice box: a long, sloping trough or box, with grooves and riffles on the bottom, into which water is directed to separate gold from gravel and sand.

Sonoma Orogeny: a period of mountain-building in western North America. This orogeny is thought to have occurred during the Permian/Triassic transition, around 250 million years ago. The Sonoma Orogeny was possibly caused by accretion of the volcanic island arc of Sonomia onto the North American continent.

South American Plate: the tectonic plate that includes South America and a large part of the Atlantic Ocean to the Mid-Atlantic Ridge.

sphalerite: zinc sulfide, a zinc ore mineral.

spreading center: the boundary, usually in the ocean, between two tectonic plates that are moving apart.

stamp mill: a type of mill that crushes material by pounding, either for further processing or for extraction of metallic ores.

stratovolcano: a cone-shaped volcano formed from many layers of ash and lava flows that build up around the vent, forming a classic steep-sided volcano.

stream piracy: the capture of water in one stream by another stream. This can happen for several reasons, including earthquakes, landslide damming, headward erosion, and glacial damming. The additional water flowing down the dominant stream can accelerate erosion and encourage the development of a gorge.

striations: a series of ridges, furrows, scratches, or linear marks caused by glaciers moving over a rock surface.

strike-slip fault: a fault in which the two sides move horizontally past each other. They are classified as right-lateral or left-lateral depending on whether the opposite side moves right or left with respect to the viewer.

subduction: the downward movement of the edge of a plate of Earth's crust into the mantle beneath another plate.

subduction zone: where Earth's tectonic plates dive back into the mantle. An oceanic trench shows where the plate is going down, and an inclined zone of earthquakes shows the path.

syncline: a bowl- or trough-shaped fold in layered rocks.

tailings: material left over after separating the valuable ore from the uneconomic rock.

talc: hydrous magnesium silicate, a soft white mineral. It is often the end product of alteration of other rocks.

talus: coarse rock debris that accumulates at the base of a cliff or escarpment.

tarn: a small mountain lake, usually gouged out by a glacier.

tectonic plates: large pieces of the Earth's crust and uppermost mantle; these plates include both oceanic crust and continental crust. Their margins are defined by zones of earthquakes.

tephra: fragments of rock and ash ejected during a volcanic eruption.

terminal moraine: piles of rock and debris carried by glaciers and dropped at their farthest extent.

terrane: the area or surface over which a particular rock or group of rocks is prevalent, or that has a distinct geologic history.

terrazzo: a composite floor or wall treatment consisting of chips of various materials.

terroir: the complete environment in which wine is produced, including such factors as soil, topography, and climate.

thrust fault: a fault in which the upper block, above a nearly horizontal fault plane, moves up and over the lower block. Thrust faults are common in areas of compression.

till: unsorted material deposited directly by a glacier and showing no layering.

tonalite: a light-colored, coarse-grained igneous intrusive rock with a composition similar to granite.

transform fault: a type of fault in which two tectonic plates slide horizontally past one another.

transform margin: a plate boundary or transform fault where the motion is dominantly horizontal.

travertine: a sedimentary rock formed by the chemical precipitation of calcium carbonate from fresh water in hot springs. Compositionally it is the same as limestone.

tremolite: a white to green silicate mineral formed by metamorphism of sediments rich in dolomite and quartz. Nephrite, one of the two minerals known as jade, is a green variety of tremolite. Fibrous tremolite is one of the six forms of asbestos.

tsunami: a giant wave caused by earthquakes or volcanic eruptions under the sea.

tufa: a variety of limestone formed when carbonate minerals precipitate out of cool water in rivers or lakes.

tuff: a relatively soft, porous rock that is usually formed by compaction and cementation of volcanic ash.

turbidite: a deposit of turbidity currents; it commonly shows bedding that grades from coarse at the base to fine-grained at the top.

turbidite fan deposit: a fan-shaped deposit in the deep ocean that is a result of a turbidity current.

turbidity current: a submarine flow of sediment suspended in water moving down a slope, usually a continental slope, into the deep ocean.

Turritella: a genus of medium-sized sea snail with a spiral cone-shaped shell.

turtleback structure: a dome-shaped range front created by undulations along exposed low-angle fault surfaces. Beneath the surface is metamorphic basement; above it are younger metamorphic and sedimentary units.

ulexite: a white borate mineral known as "cottonball" because it occurs as cottonball-sized white nodules. It is usually found in deposits left over from the evaporation of lake water.

unconformity: a break in the rock record that indicates a time gap and erosion.

vein: a fracture in rock filled by minerals or ore.

volcanic arc: also called a magmatic arc, this is a belt of volcanoes formed above a subducting oceanic tectonic plate, with the volcanoes arranged in an arc shape.

volcanic ash: a mixture of rocks, minerals, and glass particles expelled from a volcano during an eruption.

volcanic crater: a bowl-shaped depression that usually lies directly above the vent from which volcanic material is ejected.

volcanic plug: a landform created when magma hardens inside the vent of a volcano.

wave-cut terrace: a beach or platform created as waves erode the rocks along the shore into a nearly flat surface.

welded tuff: lava and ash particles ejected during eruption that were still very hot when they fell to earth. The larger lumps become flattened, and all the particles become welded together.

Notes

PREFACE
1 Prost, *North America's Natural Wonders*.
2 Alden, *Deep Oakland*, xiii.

A VERY BRIEF INTRODUCTION TO CALIFORNIA GEOLOGY
1 Anderson, *Tending the Wild*, xxviii, 34–37.

CHAPTER 1: COAST RANGES
1 There are various stories explaining the origin of the "Sleeping Lady" nickname. See, e.g., Heilig, "Tom Killion."
2 Mikkelson, "Ronald Reagan."
3 Golden Gate Bridge Highway and Transportation District, "What's in a Name—The Golden Gate?" https://www.goldengate.org/bridge/history-research/statistics-data/whats-in-a-name/#:~:text=Fremont%20gazed%20at%20the%20narrow,%E2%80%9D%20or%20%E2%80%9CGolden%20Gate%E2%80%9D%20for.
4 Gilbert, *Hydraulic-Mining Débris*, 37.
5 Gilbert et al., "San Francisco Earthquake," 9.
6 Winchester, *Crack*. (No page number available.)
7 "A Brief History of Wine in California's Napa Valley," Cult Wine Investment, September 30, 2020. https://www.wineinvestment.com/us/learn/magazine/2020/09/a-brief-history-of-wine-in-californias-napa-valley/.

CHAPTER 2: THE GREAT VALLEY
1 Stearns, "Certain Australian Forest Trees," 8.
2 Brewer, "Tamalpais and Diablo," entry for Wednesday, May 7, 1862.

CHAPTER 3: GOLD RUSH COUNTRY
1 Akins and Bauer Jr., *We Are the Land*, 2–3.
2 Covert, "Continuing Hold," 20.
3 Lang, "The Great California Gold Rush?" 14.
4 Quoted in Ewis, "Ho! For Wute Yomi!" 12.
5 Akins and Bauer Jr., *We Are the Land*, 115–16.
6 Elias, "California Gold Rush Education," 31.
7 "History of the Hangtown Fry and Recipes," City of Placerville, n.d. https://www.cityofplacerville.org/history-of-the-hangtown-fry-and-recipes.

CHAPTER 4: SIERRA NEVADA
1 Muir, *My First Summer*, 314–16.
2 Muir, "Glaciers in Yosemite."
3 Stock et al., "Impending Loss."

4 Anderson, *Tending the Wild*, 19.
5 Quoted in Russell, *Grizzly Country*, 106.
6 Russell, 107.
7 Russell, 108–109.
8 Quoted in Anderson, 19.
9 Russell, 109.
10 Russell, 112–13.
11 Wikipedia, s.v. "California Grizzly Bear," last modified October 7, 2024, 15:52, https://en.wikipedia.org/wiki/California_grizzly_bear.
12 Wikipedia, s.v. "California Grizzly Bear."
13 "Old Print Article: 'Seth Kinman, The Pacific Coast Nimrod Who Gives Chairs to Presidents,' New York Times (1885)," Afflictor.com, December 19, 2010, https://afflictor.com/2010/12/19/old-print-article-seth-kinman-the-pacific-coast-nimrod-who-gives-chairs-to-presidents-new-york-times-1885/.

CHAPTER 5: BASIN-AND-RANGE

1 Twain, *Roughing It*.
2 Wilkerson et al., *Roadside Geology*, 28.
3 Piatt, quoted on the website Bodie, California: History and Research, https://www.bodiehistory.com/bodie.htm.
4 "Bodie State Historic Park: A Gold Mining Icon in Arrested Decay," *The Diggings*, June 16, 2018, https://news.thediggings.com/bodie-state-historic-park/.
5 Piatt.
6 Twain, *Life on the Mississippi*, 58.
7 Whitney, *Yosemite Book*, 98.
8 Holt, "Paiute Indians of Utah"; Akins and Bauer Jr., *We Are the Land*, 22.
9 Wilkerson et al., 58–61.
10 Wilkerson et al., 138–40.
11 Stringfellow, "Timbisha Shoshone."
12 Cotton, *To Make a Better Nation*, vii; Stringfellow, 3.
13 The attribution to Mark Twain may be apocryphal, in that others have attributed a version of this quote to the architect Frank Lloyd Wright: "This country is built on a tilt, and sooner or later, everything that isn't nailed down ends up in California."

Selected Bibliography

The scientific information in this book draws on my experience as a professional geologist. The notes and bibliography for this book focus on historical information that is less likely to be common knowledge.

Akins, D. B., and W. J. Bauer Jr. *We Are the Land: A History of Native California.* Oakland: University of California Press, 2021.

Alden, A. *Deep Oakland: How Geology Shaped a City.* Berkeley, CA: Heyday, 2023.

Anderson, M. K. *Tending the Wild: Native American Knowledge and the Management of California's Natural Resources.* Berkeley: University of California Press, 2005.

Brewer, W. H. "Tamalpais and Diablo." In *Up and Down California in 1860–1864; The Journal of William H. Brewer.* New Haven, CT: Yale University Press, 1930. Accessed from Yosemite Online. https://www.yosemite.ca.us/library/up_and_down_california/3-2.html.

Cotton, T. *To Make a Better Nation: An Administrative History of the Timbisha Shoshone Homeland Act.* Report prepared for Death Valley National Park, 2009. https://npshistory.com/publications/deva/better_nation.pdf.

Covert, S. "The Continuing Hold of the Gold Rush on California's Indigenous Survivors." *News from Native California* 30, no. 4 (2017): 15–16.

Elias, I. "California Gold Rush Education." *News from Native California* 30, no. 4 (2017): 27–33.

Ewis, Y. "Ho! For Wute Yomi! Land of the Dead." *News from Native California* 30, no. 4 (2017): 12.

Gilbert, G. K., R. L. Humphrey, J. S. Sewell, and F. Soulé. *The San Francisco Earthquake and Fire of April 18, 1906: And Their Effects on Structures and Structural Materials*. US Geological Survey Bulletin 324, 1907. https://pubs.usgs.gov/bul/0324/report.pdf.

Gilbert, G. K. *Hydraulic-Mining Débris in the Sierra Nevada*. US Geological Survey Professional Paper 105, 1917. https://doi.org/10.3133/pp105.

Heilig, S. "Tom Killion on Mount Tam." *West Marin Review* 3 (Fall 2010). https://westmarinreview.org/volume-3/steve-heilig/.

Hill, D. P., R. H. Bailey, M. L. Sorey, J. W. Hendley II, and P. H. Stauffer. "Living with a Restless Caldera—Long Valley, California." US Geological Survey Fact Sheet 108-196, online version 2.1, 2000. https://pubs.usgs.gov/dds/dds-81/Intro/facts-sheet/fs108-96.html.

Hollett, K. J., W. R. Danskin, W. F. McCaffrey, and C. L. Walti. *Geology and Water Resources of the Owens Valley, California*. US Geological Survey Water Supply Paper 2370-B, 1991. https://doi.org/10.3133/wsp2370B.

Holt, R. L. "Paiute Indians of Utah." *Utah History Encyclopedia*. Accessed July 14, 2024. https://www.uen.org/utah_history_encyclopedia/p/PAIUTES.shtml.

Houston, J. W., and J. D. Houston. *Farewell to Manzanar: A True Story of Japanese American Experience during and after the World War II Internment*. New York: Houghton Mifflin Harcourt, 1973.

Huber, N. K. *The Geologic Story of Yosemite National Park*. US Geological Survey Bulletin 1595, 1987. https://pubs.usgs.gov/bul/1595/report.pdf.

Lang, D., and P. Jones. "Death of a Valley." *Aperture* 8, no. 3 (1960). https://archive.aperture.org/article/1960/3/3/death-of-a-valley-2.

Lang, J. "The Great California Gold Rush?" *News from Native*

California 30, no. 4 (2017): 13–14.

Matthes, F. E. *Geologic History of the Yosemite Valley*. US Geological Survey Professional Paper 160, 1930. https://doi.org/10.3133/pp160.

McPhee, J. *Assembling California*. New York: Farrar, Straus and Giroux, 1994.

Meldahl, K. H. *Hard Road West: History and Geology along the Gold Rush Trail*. Chicago: University of Chicago Press, 2007.

Mikkelson, D. "Ronald Reagan 'If You've Seen One Tree...'" *Snopes*, June 7, 2006, https://www.snopes.com/fact-check/if-youve-seen-one-tree/.

Muir, J. "Glaciers in Yosemite." *New York Tribune*, December 5, 1871. https://vault.sierraclub.org/john_muir_exhibit/writings/yosemite_glaciers.aspx.

Muir, J. *My First Summer in the Sierra*. Boston: Houghton Mifflin, 1911. https://www.yosemite.ca.us/john_muir_writings/my_first_summer_in_the_sierra/my_first_summer_in_the_sierra.pdf.

Piatt, M. H. *"The Mines Are Looking Well...": The History of the Bodie Mining District, Mono County, California*. El Sobrante, CA: North Bay Books.

Prost, G. L. *North America's Natural Wonders*. Vol. 1, *Canadian Rockies, California, The Southwest, Great Basin, Tetons-Yellowstone Country*. Geologic Tours of the World. Boca Raton, FL: CRC Press, 2020.

Russell, A. *Grizzly Country*. New York: Nick Lyons Books, 1967.

Stearns, R. E. C. "On the Economic Value of Certain Australian Forest Trees and Their Cultivation in California." *Proceedings of the California Academy of Science*, July 1, 1872. Accessible through Internet Archive: https://archive.org/stream/oneconomicvalue00stea/oneconomicvalue00stea_djvu.txt.

Stock, G. M., R. S. Anderson, T. H. Painter, B. Henn, and J. D. Lundquist. "Impending Loss of Little Ice Age Glaciers in Yosemite National Park." Paper 361-6. Geological Society of America Annual Meeting, Seattle, WA, 2017. https://gsa.confex.com/gsa/2017AM/webprogram/Paper299617.html.

Stringfellow, K. "How the Timbisha Shoshone Got Their Land Back." *PBS SoCal*, July 8, 2016. https://www.pbssocal.org/shows/artbound/how-the-timbisha-shoshone-got-their-land-back.

Twain, M. *Roughing It*. Hartford, CT: American Publishing Company, 1872. https://www.gutenberg.org/files/3177/3177-h/3177-h.htm.

Twain, M. *Life on the Mississippi*. Boston: James R. Osgood and Company, 1883.

Whitney, J. D., Jr. *The Yosemite Book: A Description of the Yosemite Valley and the Adjacent Region of the Sierra Nevada, and of the Big Trees of California, illustrated by maps and photographs*. New York: Julius Bien, 1868.

Wilkerson, G., M. Milliken, P. Saint-Amand, and D. Saint-Amand. *Roadside Geology and Mining History: Owens Valley and Mono Basin*. Washington, D.C., and Bakersfield, CA: US Bureau of Land Management and Buena Vista Museum of Natural History, 2019. https://www.academia.edu/33103580/OWENS_VALLEY_AND_MONO_BASIN_ROADSIDE_GEOLOGY_AND_MINING_HISTORY_TEXT.

Winchester, S. *A Crack in the Edge of the World: America and the Great California Earthquake of 1906*. New York: HarperCollins, 2005.

About the Author

GARY L. PROST has been working as a geologist for more than fifty years, specializing in mineral and oil exploration. He studied geology at the Colorado School of Mines, obtaining his PhD there in 1986. His employers have included the US Geological Survey and several oil companies, and his work has taken him to more than thirty countries. Now retired, he leads field trips and does public outreach on topics of geological interest. His books include *The Geology Companion*, *Remote Sensing for Geoscientists*, *English-Spanish and Spanish-English Glossary of Geoscience Terms*, *The United Kingdom's Natural Wonders*, *South America's Natural Wonders*, and the two-volume *North America's Natural Wonders*. He lives in El Cerrito, California.

A Note on Type

This book is set in Adobe Jenson Pro. Created by Adobe designer Robert Slimbach and published in 1995, it is a contemporary interpretation of the namesake typeface cut by the Venetian type designer Nicolas Jenson in 1469. For the italics, Slimbach drew on the styles of sixteenth-century designer Ludovico Vicentino degli Arrighi as revived by Frederic Warde in the early twentieth century. Adobe Jenson Pro has received wide praise for its graceful fusion of Renaissance type styles. Some of the headers in this book are set in Fira Sans, a typeface designed by the studio Carrois in collaboration with Erik Spiekermann.